Functions of Mathematics

for the

Liberal Arts

Second Edition (July 11, 2022)

by

Craig M. Johnson, Ph.D.

Professor of Mathematics

Marywood University

Scranton, PA

Functions of Mathematics for the Liberal Arts

Preface

This textbook is intended for a one or two-semester course in mathematics for college students majoring in education, English, history, music, art, foreign language, philosophy, or any other liberal arts discipline. It showcases modern applications of mathematics in many areas as well as the aesthetic features of this very rich facet of the history and ongoing advancement of human society. It works well for the type of course that is now a graduation requirement at a majority of American universities for students who are not planning a career in a field related to engineering, business, or the sciences. The prerequisite for a taking a course that uses this book is intended to be competency in intermediate algebra although most algebra skills needed beyond rudimentary computation are reviewed in the many examples that have been worked out explicitly in each section.

Philosophy

I have often heard prominent mathematicians express the sentiment that *functions*, more than any other single topic, is a central concept that threads through most areas of mathematics. Therefore, the concept of function has been chosen as a unifying theme for the entire book. The first chapter sets the stage for the rest of the text by exploring the true nature of a function via presenting a wide variety of examples, demonstrating the pattern recognition used in the creation of a function, and studying how the graph of a function aids in examining its behavior. Early development of this concept helps strengthen the ability of the student to understand later topics at a deeper level.

I believe that both the use of the concept of function as a central theme and the inclusion of several chapters seldom seen in other texts of this genre considerably broadens the appeal of this book. It provides a sample of the next level of mathematics for which a college student who has passed intermediate algebra should be able to master. The book attempts to answer the questions, "How does mathematics help us to better our society and understand the world around us?" and "What are some of the unifying ideas of mathematics?" The central theme helps to impress upon the student the feeling that mathematics is more than a disconnected potpourri of rules and tricks. Although it would be inappropriate to force a functional connection in every single section, the theme is used whenever possible to provide conceptual bridges between chapters. Developing the concept of a function augments the presentation of many topics in every chapter, including:

- The study of growth models and optimization (Chapter 1).
- Many of the main ideas in astronomy such as distance calculation, velocity, acceleration, force, and planetary models (Chapter 2).
- A clear introduction to the use of algorithms for computer programming (Chapter 3).
- Computation of future value of annuities and how loans are amortized (Chapter 4).
- The search for optimal circuits in a graph (Chapter 5).
- The dependence of election outcomes on the voting method (Chapter 6).
- The notion of geometric transformations and number-theoretic functions in music theory and cryptology (Chapter 7).
- The study of probability distributions (Chapter 8).
- The construction of confidence intervals in statistics (Chapter 9).

Objectives

This text was designed to implement NCTM (National Council of Teachers of Mathematics) curriculum standards. The topics throughout the text were chosen specifically to:

- strengthen estimation and computational skills
- utilize algebra concepts
- emphasize problem-solving and reasoning
- emphasize pattern and relationship recognition
- highlight importance of units in measurement
- highlight importance of the notion of a mathematical function
- display mathematical connections to other disciplines.

However, it must be emphasized that the main idea of this book is to provide the student with an exposure to *both* the widespread application power of mathematics as well as a perspective on its esthetically pleasing characteristics. My experience has been that a full diet of the strictly "practical approach" often appears to the student as a dreary unbroken series of formulas to be memorized and soon forgotten.

Features

• *Writing style*

The approach and writing style of this text is informal and colloquial. Much of the exposition is example-driven. A formal definition is given as a stand-alone statement only when it is essential in crystallizing the concept for the student. The goal is for the student to see where the mathematics "fits", how it encapsulates knowledge, and why its ability to predict is so powerful. Hopefully problems will be tackled not just because they are assigned, but because the answer seems intriguing and relevant. Situations have been presented to peak curiosity, inspire investigation without a sense of drudgery or, at the very least, to occasionally raise an eyebrow and prompt the comment "No kidding?"

• *Abundance of examples*

In each chapter examples of particular concepts often precede the presentation of the mathematical formalism in order to provide motivation for how situations are modeled mathematically. They have been carefully chosen to demonstrate precisely how mathematics is used to both find answers to practical problems and to provide a language that describes the behavior of natural phenomena. The units that accompany answers are stressed. The problem sets are arranged in parallel with the development in the text and ordered by increasing complexity. Many applications concern real-world problems with real data.

• *Nonstandard material*

Several chapters are included that are seldom seen in other books of this genre. Astronomy, in particular, serves well as a vehicle to showcase the historical development of science, its attendant methodologies, and the key role that mathematics plays as the language in which new science is created. Additionally, it offers an opportunity to discuss some key moments in the lives of some of the main contributors (Copernicus, Kepler, Galileo, Newton) and reveal some very human characteristics of these towering figures. The student therefore also learns that new mathematics and science do not appear suddenly in isolated bursts but are much affected by the humans who pioneer them and the culture in which they arise.

The chapter on connections of math to music also is a novel topic enjoyed by many of the students who populate this type of class. The various uses of number theory and geometry are explored in identifying and creating musical patterns and in musical composition. Since most people never associate applications of mathematics to disciplines other than the mainstream examples in engineering and science, many students are delighted with this chapter. I have witnessed many grins and heard many favorable comments from art and theater majors about how much they enjoyed this material. And again, the discussion of the musicians themselves (Bach, Brahms, Mozart, Joel) brings home a human element that cannot be so readily examined in the other more standard topics.

• *Vocabulary list*

Every discipline has its own vernacular that is essential for accurate explanations of the central ideas. A list of all the key new vocabulary words is given at the end of each chapter along with their definitions.

• *Chapter Review Tests*

Students appreciate knowing the types of problems they should be expected to know how to solve and the format in which they will be given. It has been shown that good pedagogy includes providing clear expectations. The chapter review tests give students a framework to help them properly prepare and eases the anxiety that can arise from "fear of the unknown".

• *Historical Figures*

Brief accounts of the lives of the key mathematical contributors are presented in most chapters directly imbedded in the text rather than boxed and set to the side in the style used in most books. The idea is to emphasize that mathematical development definitely occurs *within* the human experience rather than up in the attic apart from one's daily life. I strongly feel that the student is well-served by a healthy dose of appreciation for the historical development of mathematics within the human experience and its associated impact on how we view the world.

• Flexibility

A semester course usually consists of three or four chapters selected according to student need or interest. Coverage of four chapters is usually only possible if at least two of the chapters consist of three rather than four sections. A good rule of thumb is to limit the total number of sections to less than or equal to fourteen. If *Logic* or *Music* is included in a four chapter semester, the fourth section can be omitted if time does not allow. Over the past fifteen years I have used many different groupings including *Functions, Astronomy*, and *Music*; *Functions, Graph Theory, Voting Methods*, and *Music*; *Graph Theory, Voting Methods, Probability*, and *Statistics*; *Functions, Logic*, and *Personal Finance*; *Astronomy, Logic*(excluding the fourth section), *Voting Methods*, and *Music*; and many others. My habit has been to usually begin with *Functions*, but it is certainly not essential. The function theme can be expanded upon by the instructor when the concept emerges in a particular chapter.

Calculators

Calculators are assumed to be available to the student for computation throughout the textbook. If the number of significant digits used in answers is not indicated, it is assumed to be three or four, depending on the given data. One exception is the chapter on *Personal Finance*, where all answers are rounded to the nearest cent. A few exercises in the *Functions* chapter explore graphing calculator capabilities.

Acknowledgements

I wish to thank the following people for their help and valuable suggestions.

David Bock, Author of *Intro Stats, Stats: Modeling the World, AP Calculus*, and others

Victor Cifarelli, *University of North Carolina at Charlotte*

Owen Gingerich, *Harvard University*

Leon Harkelroad, *Bowdoin College*

Thomas Kent, *Marywood University*

Anthony Pusateri, *Marywood University*

Ross Rueger, *College of the Sequoias*

Beverly West, *Cornell University*

Chaogui Zhang, *Marywood University*

Dedication

To my wife Viv and my sons Nick and Scott,

whose support and witty repartee made this book possible.

Table of Contents

Chapter 1

The Concept of Function

Form and function should be one, joined in a spiritual union.   Frank Lloyd Wright

It has been argued by some educators that the concept of *function* is the single most important mathematical idea from kindergarten to graduate school. To begin by presenting a dictionary-style definition, however, may not offer any true insight to the heart of the matter. A subtler approach is needed. When teaching the meaning of a new word, an English teacher often gives a definition immediately followed by the use of the word in a sentence. Most people learn new concepts not by memorization of a formal definition but rather by a set of clear examples coupled with their personal experience connected with those examples. Understanding the meaning of a definition comes later, followed later still by perhaps more subtle ramifications of that definition. The learning of mathematical concepts usually follows a similar path. Therefore, we begin this chapter with a section that presents a collection of "experience" examples of functions – apparent relationships between sets of objects that may or may not be expressible in algebraic form. These examples illustrate the notion of a strict correspondence between sets, an idea that leads in the next section to the examination of a function between sets of real numbers as a "process" or "action". The last two sections explore the representation of a real-valued function as a graph and the powerful tool this provides in analyzing a function's behavior.

1.1 Functions Around Us

In the study of elementary algebra, letters and other symbols are introduced as convenient representatives for numbers. The first demonstrated use of a letter in mathematics is typically as a "stand-in" or a "holder" for a value which is unknown until the completion of some solution process, e.g. $2x + 7 = 19$. There is a fundamental distinction between the use of the letter x in an equation as a "holder" and its use as a *variable* quantity whose value is associated with a second variable quantity at all times, e.g. $y = 2x + 7$. In the first case, by either observation or some formal technique, we quickly reach the conclusion that the value of x is 6, period. The second case is very different. Here we are free to assign to x whatever value we like and this choice then dictates a corresponding value for y. When x is 2, y is 11, when x is 3, y is 13, and so on. It is this second case that helps us get to the definition of a function. Consider this example.

Example 1
Phillip has accepted a salesman position at a computer store and is offered a choice of compensation: either $400 per week plus 12% commission on his sales or $450 per week plus 10% commission on his sales. In order to reach a decision Phillip asks himself two questions:

    1. For what dollar amount of sales will these options yield the same payment?

    2. If Phillip feels he will consistently sell $3500 of computer equipment per week, which option of compensation will pay him the most money?

Solution The first question is definitely asking for a single numerical answer. It can be solved by the use of a single symbol. If s is the amount of sales that will yield identical payments according to each option, then s must satisfy the equation

$$\begin{aligned} 0.12s + 400 &= 0.10s + 450 \\ 0.12s &= 0.10s + 50 \\ 0.02s &= 50 \\ s &= 2500 \end{aligned}$$

Phillip would be paid the same amount by both options if he sells $2500 worth of computer equipment per week.

The second question is *not* concerned with a single number response. A comparison needs to be made between the two amounts which Phillip would earn assuming he made sales of $3500 per week. We already know these amounts *must* be different because they are the same only for sales of $2500. Each amount is to be computed according to a different option. In other words, a *relationship* needs to be written down in a "formula-type" format for each scheme which

Figure 1.1.1 Should Phil sell these?

will compute the payment based on a particular sales value.

Let y represent the amount of payment Phillip receives based on x dollars worth of sales.

Under the first option, $y = 0.12x + 400$
Under the second option, $y = 0.10x + 450$

Now we simply replace x by 3500 in each formula and calculate the corresponding amounts.

Under the first option, $y = 0.12(3500) + 400 = 820$.
Under the second option, $y = 0.10(3500) + 450 = 800$.

We see that Phillip will make more money under the first option if he sells $3500 worth of computers per week. ◆

Answering the second question required a significantly different approach, one requiring the identification of related *varying* quantities. A letter or symbol that may represent any one of a variety of different quantities is called a **variable**. Two variables were needed to form each relationship in the above example indicating that, as one quantity changed, the other quantity also changed in a well-defined way. Usually it is helpful to think of one variable freely assuming any of a host of different values and so we call it the **independent variable**. Once the independent variable is assigned a particular value, the other variable immediately obtains its own value as a direct consequence and hence is known as the **dependent variable**. In the relationship above given by $y = 0.12x + 400$, x is the independent variable and y is the dependent variable. If Phillip is paid according to this scheme, his payment y depends strictly on his sales x for the week, which would be changing from week to week. In essence, the construction of the two-variable relationship saves us the time and energy that would be needed to repeatedly solve for an unknown quantity.

Ok, so what exactly is a *function*? In common everyday conversation, this word gets used according to various forms of its definition. One might say, for example, that "The main function of the U.S. Supreme Court is to interpret the Constitution" meaning that interpretation is the characteristic action of the Supreme Court. A weatherman might say, "This drought is a function of the atmospheric pressure patterns," indicating that a connection exists between atmospheric

www.shutterstock.com · 54183202

pressure and the lack of rain. The word "result" would be just as appropriate to use as "function" here because the implied connection is a rather loose one. On the other hand, a statement such as "In our county, voter turnout is always a function of temperature", a more direct tie could exist. It may even be possible that the connection is one which could be approximately modeled by an equation, e.g. $v = 100t + 1500$, where v is the number of voters and t is any temperature between 10 and 80 degrees Fahrenheit. If t is only 30°, then 4500 people vote, but if it is a more balmy 70°, then 8500 show up at the polls. This is **Figure 1.1.2** Do more people vote on a warm day?
a relationship between an independent variable t and a dependent variable v. The set of temperatures (real numbers between 0 and 80) for which this equation holds is referred

to as its **domain** and the resulting collection of numbers of voters is called its **range**. Since mathematics is the language used to accurately record and describe many types of knowledge, mathematicians cannot afford to be ambiguous in their definitions. The function is a concept so central to the realm of mathematics that it will be one of the guiding themes of this book.

A **function** is a relationship between two groups of objects (usually numbers) such that every object in the first group, called the **domain**, is unambiguously matched up with exactly one object in the second group, called the **range**.

Domains and ranges are generally designated by the use of set or interval notation. (See Appendix A for a description of interval notation.) A **set** is a collection of objects, usually possessing some similar characteristic such as a set of cities, a set of cars, or a set of numbers. Brackets { } around a listing of the members of the set separated by commas is the standard method of representation. For instance, the set of capital letters from A to F is {A, B, C, D, E, F}. Very large sets need some abbreviation. The set of capital letters from A to Z can be expressed {A, B, C, . . ., Z}. Infinite sets trail off with . . . after establishing a pattern. For instance, the even positive integers can be denoted by {2, 4, 6, 8, . . .}.

The world around us is brimming with functions, many of which are specific enough to be represented with independent and dependent variables as in the last section. We shall call such an equation a **formula** and the important skill of creating a formula for a function by identifying the *action* it performs on elements in its domain will be studied in the next section. For the present, however, we concern ourselves with examples of naturally occurring functions, many of which cannot be translated to a neat concise formula, but are functions nonetheless. In this section, we simply want to identify the function in each example along with its domain and range.

Example 2

Suppose you are taking a walk with a pedometer on your wrist that keeps track of your travel time and distance covered and computes your walking speed. Then the distance d you travel as measured from some starting point is a function of the elapsed time t. If your current walking speed is 3.4 miles per hour, then for $t = 1$ hour, $d = 3.4$ miles, for $t = 2$ hours, $d = 6.8$ miles, and so forth. If t is measured in hours and d is measured in miles, then we can write

Figure 1.1.3 A pedometer.

$$d = 3.4\,t$$

to express this function. If your longest walk takes 3 hours, then a reasonable domain would be the interval $\mathbf{D} = [0, 3] = \{\,t$ any real number $\mid 0 \le t \le 3\,\}$ and the associated range would be $\mathbf{R} = [0, 10.2]$. (See Appendix A for a definition of intervals.) ◆

Example 3

Suppose we form a set $\mathbf{D}$ consisting of the people in your mathematics class (all having different first names) and another set $\mathbf{R}$ containing all their heights (in inches). Since every person has one and only one height, we can define a function f by assigning to each person the number corresponding to his or her height. Then $\mathbf{D}$ would be the domain of f and $\mathbf{R}$ would be the range. Figure 1.1.4 shows a schematic diagram displaying some sample assignments of this function. We think of the function as an assignment "box". One domain element into the box yields one range element out of the box.

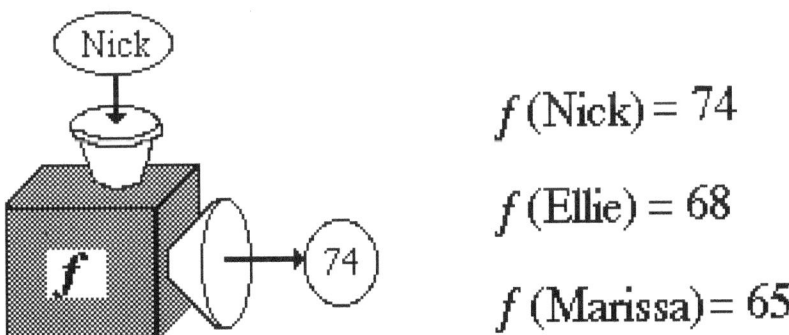

Figure 1.1.4 Since every person has a unique height, these assignments are representative of a function.

Suppose Nick is 74 inches tall. To display that specific assignment of a domain member to his corresponding range value, the conventional notation is to write: $f(\text{Nick}) = 74$. Similarly, if Ellie and Marissa are 68 and 65 inches tall respectively, we would write $f(\text{Ellie}) = 68$ and $f(\text{Marissa}) = 65$, and continue in this manner. Note, incidentally, that it is impossible to define a function having a domain of heights and a range consisting of your classmates. Two or more people, say Claudia and Rafer, could have the same height, say 70, in which case the assignment of a range member to 70 would require an arbitrary choice. If g were going to represent a possible function, $g(70)$ would not be unique and so our definition of a function would not be satisfied. ♦

Example 4 (Chemistry)

Pythagoras (c.580–500 BC.) was an early Greek mathematician and philosopher who felt that numbers held the key to any deep understanding of the world. He would have been thrilled with the *Periodic Table of the Elements* – a categorical listing of the known elements that is essential to the study of chemistry. The nucleus of an atom of each element has a unique number of protons known as the *atomic number* of the element. The numbers listed above the element symbols in the Periodic Table in Figure 1.1.5 are atomic numbers.

Figure 1.1.5 The Periodic Table of Elements.

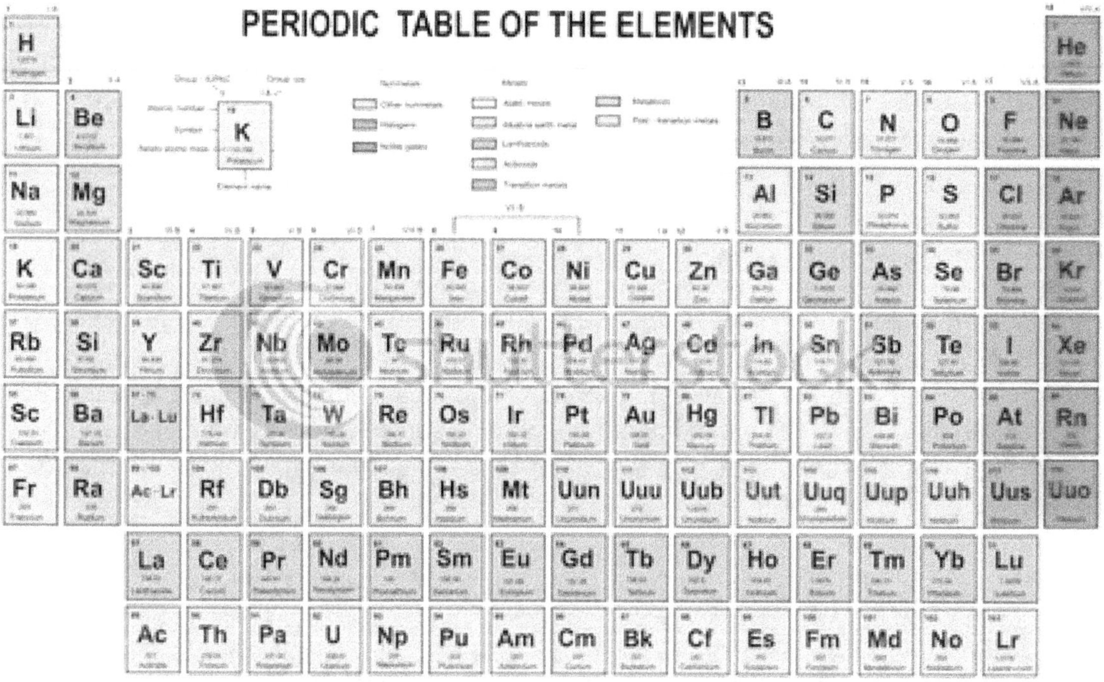

The assignment of its atomic number to each element is a functional relationship at its heart because the basic definition of what constitutes an element starts, in fact, with its atomic number and so it is unique. The first three assignments would be:

$$
\begin{array}{lll}
\text{H} & \rightarrow & 1 \qquad \text{(Hydrogen)} \\
\text{He} & \rightarrow & 2 \qquad \text{(Helium)} \\
\text{Li} & \rightarrow & 3 \qquad \text{(Lithium)}
\end{array}
$$

Borrowing the practice from algebra of using letters such as f and h for representing functions, we use f for our function and write these pairings as $f(\text{H}) = 1, f(\text{He}) = 2, f(\text{Li}) = 3$, and so on. This notation is commonly used because it is less cumbersome than above. The domain is the set $\mathbf{D} = \{\text{H, He, Li, Be, B, C}, \ldots, \text{Uuo}\}$ and the range is $\mathbf{R} = \{1, 2, 3, 4, 5, 6, \ldots, 114\}$. ♦

14

A useful analogy to use when thinking about functions is that of a bow shooting arrows onto a target. The quiver of arrows is the domain, the bow is the function itself, and the points on the target where the arrows are shot represent the members of the range. The key to this picture is to realize that the bow (function) provides the connection from quiver (domain) to target points (range) as in Figure 1.1.6.

The word *image* is also used when speaking of a specific member of the range. In the prior example, 1 is the image of H with respect to this function, 2 is the image of He, 3 is the image of Li and so on. For any function denoted by f, we shall always think of $f(x)$ as the image of x after application of that function. As the independent variable x varies in the domain, its image $f(x)$ varies throughout the range.

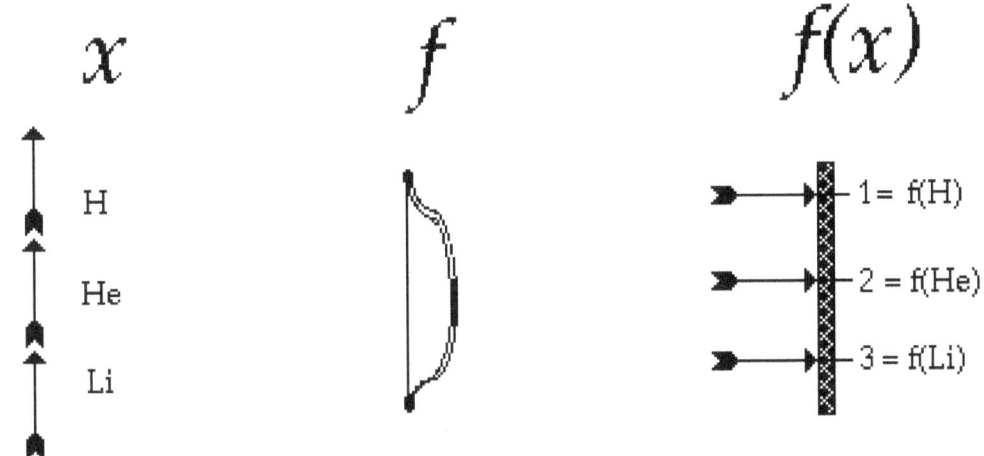

Figure 1.1.6 The function represented as a bow shooting arrows (members of the domain) onto target points (members of the range).

Example 5

In contrast to the above example, the number of *neutrons* contained in a nucleus is *not* a function of the chemical element because of the presence of isotopes. The isotopes of an element are atoms whose nuclei contain the same number of protons but different numbers of neutrons. Although carbon, for instance, normally has six protons and six neutrons in its nucleus, about one in 10^{12} atoms contain six protons and eight neutrons and is referred to as carbon 14 (used in the dating of carbonaceous materials). A function does not exist if the domain is allowed to contain isotopes because it is not possible to unambiguously select the image of C to be 6 or 8. A similar situation exists for all the other elements since every known element possesses at least two isotopes. (Figure 1.1.7)

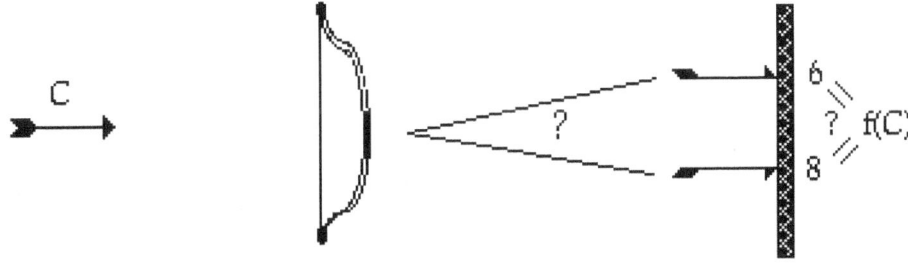

Figure 1.1.7 The assignment of an element to the number of neutrons in its nucleus is not a function.

Example 6 (Meteorology)

The high temperatures in several American cities for a typical Saturday in January are given below.

Anchorage	25	Flagstaff	58
Birmingham	45	Jacksonville	53
Casper	49	Milwaukee	21
Chicago	21	Phoenix	79
Dayton	16	Sioux Falls	30

Figure 1.1.8 High temperatures in some American cities.

Since every city can only have one high temperature for the day, we have a function between cities and temperatures. The domain here is **D** = {Anchorage, Birmingham, Casper, Chicago, Dayton, Flagstaff, Jacksonville, Milwaukee, Phoenix, Sioux Falls} and the range is **R** = {16, 21, 25, 30, 45, 49, 53, 58, 79}. Using T for the function this time (for temperature), we can write T(Anchorage) = 25, T(Birmingham) = 45, T(Casper) = 49, and so on.

The fact that both T(Chicago) = 21 and T(Milwaukee) = 21 does not keep this relationship from being a function. Two (or more) different members of the domain of any function may have the same image. However, it does prevent a function from existing in the "opposite direction", that is, with the high temperatures as the domain and the cities as the range. A city could not be uniquely identified by its high temperature. What, for instance, would be assigned to 21? The presence of an arbitrary choice for that assignment eliminates the possibility of a functional relationship. ◆

Example 7

The population of the village of Beushane has increased in recent years. In 2001 there were only 760 people. However the construction of a new highway near the village increased the population P by a constant annual amount according to the following table.

Year	Population
2001	760
2002	795
2003	830
2004	865
2005	900

At this rate of growth how many people will be living in Beushane in 2008? 2010? 2013?

Solution

If we let t (for time) be our independent variable here, it is convenient to let the initial year of 2001 correspond to $t = 0$. Then t stands for the number of years after 2001.

Year	t	Population
2001	0	760
2002	1	795
2003	2	830
2004	3	865
2005	4	900

Since the population grows by 35 people every year we see that we have a two-variable relationship given by $P = 35t + 760$. The years 2008, 2010, and 2013 correspond to t = 7, 9 and 12 years after 2001 and so the population (assuming the growth rate remains the same) would be:

$$P = 35(7) + 760 = 1005 \qquad \text{for } t = 7$$
$$P = 35(9) + 760 = 1075 \qquad \text{for } t = 9$$
$$P = 35(12) + 760 = 1180 \qquad \text{for } t = 12. \blacklozenge$$

The use of tables illustrates one of the oldest methods for representing a relationship between two varying quantities. Your familiarity with the creation of tables probably comes from experiments done in a science class or from the style with which information is often given in newspapers and magazines. Usually it is tacitly assumed that such tables are only a small subset of a much larger set of paired numbers and so we may extend the relationship implied in the table for predictive purposes. Whenever we have paired data of any type available, predictions can be made for other instances with the assumption that the relationship appearing in the listing or table holds steady. This practice is at the heart of one of the main concerns of this book and shall be explored in more detail.

Example 8 (Geometry)

The circumference C of any circle is related to its diameter d by $C = \pi d$. Since the diameter is equal to twice the radius r, this is also commonly written as $C = 2\pi r$. Clearly the area A of any circle must also depend on the size of its radius and, in fact, the specific relationship is given by $A = \pi r^2$. For instance, a circle of radius 3.00 ft has a circumference and area of

$$C = 2\pi(3.00) \approx 18.8 \text{ ft} \qquad \text{and} \qquad A = \pi(3.00^2) \approx 28.3 \text{ ft}^2.$$

It is clear that one radius leads to unique values for both C and A. So we see that the circumference and area of a circle are functions of its radius. The domains and ranges of both these functions are all given by the set of positive real numbers. In Figure 1.1.9 we note further

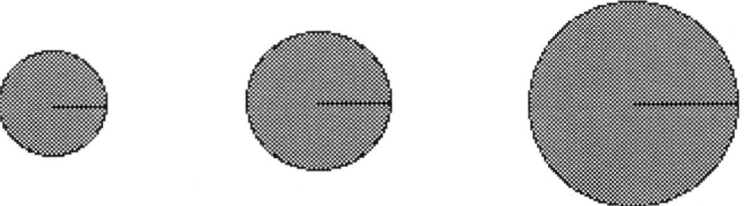

Figure 1.1.9 The area of a circle is an increasing function of radius.

that both of these physical quantities must get larger as the radius gets larger. The mathematical way to describe this property is to say that circumference and area are *increasing* functions of radius. ♦

It is possible for a function whose domain **D** and range **R** both consist entirely of numbers to have certain properties which help characterize the function. A function is called *increasing* if $f(x)$ in **R** always increases whenever x in **D** increases. One might compare this to the growth process of a child – the older she gets, the taller she gets. On the other hand, if $f(x)$ decreases whenever x increases, we call f a *decreasing* function. Any mountain climber will tell you that atmospheric pressure and air density are decreasing functions of altitude.

Example 9 (Biology)

Biologists are concerned with the concentration of hydrogen ions (H^+) and hydroxide (OH^-) in substances because of their effect on many organic processes. In a solution, the concentration of H^+ times the concentration of OH^- is always equal to the constant value 10^{-14} mole per liter (mol/L). A typical carbonated soft drink has an H^+ concentration of 10^{-3} mol/L and therefore its OH^- concentration must be 10^{-11} mol/L. Clearly each of these concentrations is a function of the other. As one increases, the other decreases in a very well-defined manner which we could actually write down as

$$H^+ = \frac{10^{-14}}{x} \quad .$$

Here we see that the hydrogen ion concentration H^+ is a decreasing function of the hydroxide concentration x. ♦

Example 10 (Zoology)

The rate at which a mammal's heart beats is approximately a decreasing function of the size of the mammal. This means that the heart-rate decreases as the size increases. Even if we lack a specific formula to make specific matches between domain and range numbers, we can still use this function to conclude that the heart-rate of a fox is greater than that of an elephant but less than that of a mouse. Knowledge of a functional property such as this can answer a great number of questions of a comparative nature. ♦

Example 11 (Astronomy)

The Greek astronomer Hipparchus (c. 2nd century BC) was the first person to catalogue over a thousand stars according to size. Since all the stars at that time were thought to be the same distance from Earth, this amounted to a classification by brightness. He created six categories: the brightest stars were of class 1, the second brightest of class 2, and so on. Latin translators of the catalogue used the word *magnitudo*, the word for size, for a class number, and so today we say, for example, that the north star, Polaris, has magnitude 2. Figure 1.1.7 displays some magnitudes (of greater accuracy because of modern instruments) for the stars in the constellation Pegasus.

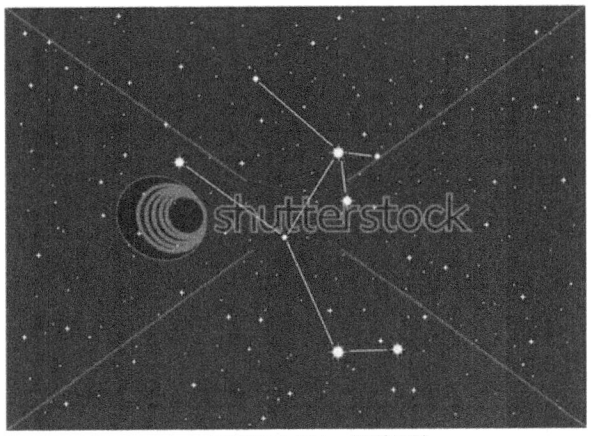

Alpheratz	2.06	Mu Pegasi	3.49
Markab	2.50	Upsilon Pegasi	4.41
Algenib	2.83	Psi Pegasi	4.64
Homam	3.39	Phi Pegasi	5.05

Figure 1.1.10 Apparent magnitudes of some stars in Pegasus.

To an astronomer in the time of Hipparchus, the size of a star was a decreasing function of magnitude and so the above table revealed that Psi Pegasi is a smaller star than Markab. Today, however, we know that the stars in the skies above us exist at many varying distances. Just as we perceive the beams differently from two similar flashlights at two different distances, so too is the apparent brightness of a star to our eyes here on Earth related not just to its size but also to its distance from us (as well as its temperature). Today the values in Figure 1.1.10 are known as *apparent* magnitudes and we may only state that apparent brightness, rather than size, is a decreasing function of apparent magnitude. In fact, Markab appears to be the brighter star in large part because its distance (73 light-years) is less than 7% the distance to Psi Pegasi (1080 light-years). ◆

Example 12 (Music)

Every musical pitch has a vibrational frequency associated with it. For instance, the "A" string on most acoustic guitars vibrates at 440 hertz (cycles/sec). This is the case regardless of the system – a string, column of air, or bar – that is being used to produce the vibrations. The values in Figure 1.1.11 correspond to one type of musical scale (chromatic) and are not exact.

Frequency Table

| | Octave | | | | |
	0	1	2	3	4
C	65.4	130.8	261.6	523.2	1046.4
D	73.4	146.8	293.6	587.2	1174.4
E	82.4	164.8	329.6	659.2	1318.4
F	87.3	174.6	349.2	698.4	1396.8
G	98.0	196.0	392.0	784.0	1568.0
A	110.0	220.0	440.0	880.0	1760.0
B	123.5	246.9	493.9	987.8	1975.6

Figure 1.1.11 Frequencies of the pitches in the chromatic scale for several octaves.

Several functions can be produced from this chart, depending on how we wish to display the relationships. For example, since each frequency produces one and only one pitch, we may write the first few pitch names as a function of various frequencies.

$f(65.4) = C$ $f(130.8) = C$ $f(261.6) = C$ . . .

$f(73.4) = D$ $f(146.8) = D$ $f(293.6) = D$ . . .

$f(82.4) = E$ $f(164.8) = E$ $f(329.6) = E$ . . .

So we have a relationship where pitch is a function of frequency. Every musician who has ever played an instrument has utilized this function. The domain here consists of **D** = { 65.4 73.4, 82.4, . . . } and the range **R** = { C, D, E, F, G, A, B }. Observe also that this function takes more than one element in the domain to the same element in the range. *This is OK* ! This does not violate the definition of a function, since every domain member does get sent to a *unique* range member. In our bow and arrow analogy, it would correspond to more than one arrow being shot to the same point on the target (Figure 1.1.12). ◆

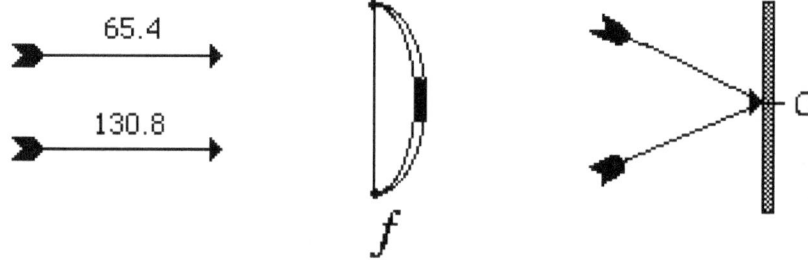

Figure 1.1.12 A function may send more than one domain element to the same range element.

Example 13 (Cryptology)

The study of integers and their related properties is an area of mathematics known as *number theory* and any function whose domain is the set of positive integers is called a *number-theoretic* function. One such function is denoted by the Greek symbol τ ("tau") and defines τ(*n*) to be the number of positive divisors of *n* , where *n* is any positive integer. Thus, τ(10) = 4, because 10 has the four positive divisors 1, 2, 5, and 10. Similarly, τ(30) = 8, because the divisors of 30 are 1, 2, 3, 5, 6, 10, 15, and 30. Other examples are given in Figure 1.1.13.

n	Positive Divisors of n	τ(n)
4	1, 2, 4	3
7	1, 7	2
8	1, 2, 4, 8	4
9	1, 3, 9	3
12	1, 2, 3, 4, 6, 12	6
15	1, 3, 5, 15	4

19	1, 19	2
40	1, 2, 4, 5, 8, 10, 20, 40	8
100	1, 2, 4, 5, 10, 20, 25, 50, 100	9
200	1, 2, 4, 5, 8, 10, 20, 25, 40, 50, 100, 200	12

Figure 1.1.13 τ(n) for various values of n.

We see that this is another example of a function that assigns the same (range) element to more than one element in its domain. In fact, it is true that that for every $k > 1$, there are an infinite number of integers which have k positive divisors.

Number theory is used extensively in **cryptology** – the study of the techniques used in the creation of secret codes. Long used primarily for military communications, applications of cryptology have recently increased dramatically as the need has arisen for security in the ever more sophisticated computer systems that contain and control large data bases. ◆

When both the domain and range of our function consist of real numbers, often the relationship can be conveniently expressed using an independent variable in an equation. This takes the form of a *formula* of the type you are accustomed to from algebra. Sometimes a dependent variable is also used in the equation to represent values in the range, but more often we continue to use the functional notation device $f(x)$ to stand for the value of the function corresponding to x. If $f(x) = 3x^2 + 2x - 5$ for instance, then

$$f(-4) = 3(-4)^2 + 2(-4) - 5 = 35$$
$$f(1) = 3(1)^2 + 2(1) - 5 = 0$$
$$f(2) = 3(2)^2 + 2(2) - 5 = 11$$

and so forth. This ideal notation was first introduced by the prolific mathematician extraordinaire **Leonhard Euler** (1707–1783), a man who did mathematics "without apparent effort, as men breathe, or as eagles sustain themselves on the wind."
Born and educated in Switzerland, Euler had received two degrees by the age of 16 at the local university. His prodigious and restless intellect was given free reign in his career as royal mathematician at the courts of Russia and Berlin to produce an amount of new mathematics so enormous that people are still examining portions of his works today. His memory was legendary – in his old age he could still recite every word of the *Aeneid* by Virgil even knowing the first and last sentence of every page! He was one of the first to utilize the full power of the calculus developed by **Isaac Newton** (1642–1727) and he

Figure 1.1.14 Leonhard Euler.

realized the importance of functions to concisely explain relationships and patterns among physical phenomena. Because one of his specialties was the creation of **algorithms** or step-by-step procedures for the solution of tough problems, the clear definition of a function was an important tool to him.

21

A function of a variable quantity is an analytic expression composed in whatever manner of this same quantity and numbers or constant quantities.

Leonhard Euler

Example 14 (Physics)

Pendulums were an important early mechanism for timekeeping and therefore an object of study by the scientist and mathematician **Galileo Galilei** (1564–1642). (See Section 3.3 for more details on his life). The time for a pendulum to complete one swing back and forth is called its period of oscillation T. Our common experience would indicate that the longer the length x of the pendulum, the greater is the value of T. In fact, for small oscillations it was determined that

$$T(x) = 2\pi\sqrt{\frac{x}{g}}$$

where g is the acceleration due to gravity, x is measured in meters, and T is in seconds. On the surface of the earth $g \approx 9.8$ m/sec^2 giving $\dfrac{2\pi}{\sqrt{9.8}} \approx 2.0$.

So the formula for the function reduces to $T(x) = 2.0\sqrt{x}$.

Figure 1.1.15 Galileo (1564-1642)

The periods for pendulums of lengths 0.25 m, 0.8 m, and 1.5 m are
$$T(0.25) = 2.0\sqrt{0.25} = 1.0 \text{ sec,}$$
$$T(0.8) = 2.0\sqrt{0.8} = 1.8 \text{ sec,}$$
$$T(1.5) = 2.0\sqrt{1.5} = 2.4 \text{ sec.}$$

The domain and range both must consist of all positive real numbers. Note that T is used here to represent the function instead of f. It is quite common to use letters for functions that remind us of the quantity being symbolized. ♦

Legends of Galileo recount an early mystical experience in church that fostered his profound insights about the pendulum as timekeeper: [He was] mesmerized by the to-and-fro of an oil lamp suspended from the nave ceiling and pushed by drafts. Timing the motion of the lamp by his own pulse, Galileo saw that the length of a pendulum determines its rate.

Dava Sobel in *Longitude*

Example 15 (Business)

If you invest a principal of P dollars in a savings account which earns *simple* interest at an annual rate of r, then the sum of money $S(t)$ you will have accrued in t years is a function of t given by

$$S(t) = P(1 + rt)$$

where r has first been converted from a percentage to its decimal equivalent. (Note that t is used for the independent variable because it represents time.) Suppose you invest a principal of \$700 in an account earning simple interest at an annual rate of 3.5%. Then the above formula becomes the increasing function

$$S(t) = 700(1 + 0.035t).$$

In 2 years, you will have accrued $\quad S(2) \quad = 700(1 + 0.035 \cdot 2) = 700(1.07) = \$749.00$
In 3 years, $\qquad\qquad\qquad\qquad\quad S(3) \quad = 700(1 + 0.035 \cdot 3) = 700(1.105) = \$773.50$
In 5 years, $\qquad\qquad\qquad\qquad\quad S(5) \quad = 700(1 + 0.035 \cdot 5) = 700(1.175) = \$822.50$
In 10 years, $\qquad\qquad\qquad\qquad\quad S(10) = 700(1 + 0.035 \cdot 10) = 700(1.35) = \$945.00$ ◆

Exercise Set 1.1

1. The low temperatures in several cities on a Saturday in January are given below. Describe a function and give the domain and range in set notation. Also give three examples of functional pairings using the "f(domain element) = range element" notation.

Amsterdam	39	Oslo	24
Calgary	14	Rome	43
Dublin	41	Singapore	75
Lisbon	43	Tokyo	36
Manila	66	Vienna	27

2. Could you define a function from a domain of the above temperatures to a range of the corresponding cities? Explain.

3. A typical program at a professional sports contest contains a list of the players with both their heights and weights. Does a function exist from a domain of weights to a range consisting of the players? Explain.

4. The rate at which a mammal's heart beats is approximately a *decreasing* function of the size of the mammal. A baboon's heart-rate is about 100 beats per minute, a lion's about 50, and an elephant's about 25. This information does not permit us to construct specific matches of mammals with heart-rates for the function, but we can still answer the following questions. In what interval do you think the heart-rate of a full-grown rhinoceros would be: 0-40, 40-80, or over 80? A zebra? A rabbit? Which has the faster heart-rate, a 10-year-old child or an adult? A 10-year-old child or an infant?

5. Suppose your teacher for a certain class posts the grades for each test next to a list of social security numbers. Use the notion of a function to explain why this is sufficient for each student to be informed of his or her grade.

6. Suppose during this year that July 21 is a Friday. Do you think July 21 is a Friday every year? Is it possible to create a single function from date to day of the week that is good for all years?

7. The multiple of 4 that is closest to 51 is 52. The closest multiple of 4 to 81 is 80. If we define a method for assigning numbers to an arbitrary integer x by
$$x \rightarrow \text{closest multiple of 4,}$$
is this a function? Explain.

8. At the end of the season the scoring average per game is computed for each player in the National Basketball Association. If you were to define a function between the set of players and the set of scoring averages, which set must be the domain and which must be the range?

9. The prices for a six-pack of soda at the local grocery store are given below. Could you define a function with domain consisting of the prices and a range consisting of the soda brands? Why or why not?

$2.58	Cran-Cola
$2.45	Ginger Ale
$2.25	Grape Surprise
$2.62	Raspberry Sluice
$2.25	Al's Root Beer
$2.08	Orange Fizz

10. The frequency ν (Greek "nu") and wavelength λ ("lambda") of any type of radiation in the electromagnetic spectrum are related by the equation $\nu = \dfrac{c}{\lambda}$ where c is the speed of light and therefore a constant value. Is ν an increasing or decreasing function of λ ?

11. When you stand close to a roaring campfire you feel a great deal of heat, but as you recede you notice a rapid lessening of the warmth. Is the warmth you feel an increasing or decreasing function of your distance from the fire?

12. In Example 13, the number-theoretic function τ was defined by:
$\tau(n)$ = number of positive divisors of n , where n is any positive integer.
Determine $\tau(6), \tau(11), \tau(28), \tau(33), \tau(49), \tau(53), \tau(64)$, and $\tau(87)$.

13. A number-theoretic function often associated with the τ function is denoted by the Greek letter σ ("sigma") and is defined by:
$\sigma(n)$ = sum of the positive divisors of n, where n is any positive integer.
For example, the divisors of 10 are 1, 2, 5, and 10. Therefore, $\sigma(10) = 1 + 2 + 5 + 10 = 18$. Use Figure 1.1.13 to find $\sigma(4), \sigma(12), \sigma(19)$, and $\sigma(100)$.

14. Use your work from Exercise #12 to help determine $\sigma(6), \sigma(11), \sigma(28), \sigma(33), \sigma(49), \sigma(53)$, $\sigma(64)$, and $\sigma(87)$.

15. If p is an integer greater than 1 whose only positive divisors are itself and 1, then p is called a *prime* number. The first few primes are 2, 3, 5, 7, 11, 13, 17, 19, 23, 29, 31, etc. Find expressions for $\tau(p)$ and $\sigma(p)$ if p is a prime.

Find the images of each of the following functions at the given values for the independent variable.

16. $f(x) = 9x + 2, \; x = -2, 0, 3, 10, 70$

17. $f(x) = x^2 + 4x - 3, \; x = -6, -1, 0, 2, 10$

18. $g(x) = 5x + 3\sqrt{x} \; , \; x = 4, 9, 25$

19. $h(t) = \dfrac{2t + 7}{10} \; , t = -1, 1, 8, 25, 100$

20. $A(r) = \pi r^2 \, , \; r = 2, 7, 15, 20$

21. $D(y) = 6\sqrt{y^2 + 7} \; , y = -3, -2, 0, 5$

22. $F(x) = 5x^3 - 2x^2 - 0.4x + 3, \; x = 1.2, 3.7, 4.5$

23. $G(t) = \qquad , t = 0, 1, 2, 3$

24. Recall from Example 14 that the period of oscillation T (in seconds) for a pendulum is a function of its length x (in meters) given by

$$T(x) = 2.0\sqrt{x}$$

Determine $T(0.2)$, $T(0.8)$, and $T(1.3)$. Round your answers to the nearest tenth. Is T an increasing or decreasing function of x?

25. In Example 15 we saw that the amount of money accrued in t years by investing P dollars in a savings account earning simple interest at an annual rate r is given by $S(t) = P(1 + rt)$. For an investment of \$7300 at an annual rate of 4.5%, find:
     i) the specific function giving the amount of money accrued in t years.
     ii) the amount of money accrued in 3 years.
     iii) the amount of money accrued in 5 years.

26. Deion invests \$325,000 in a savings account earning simple interest at an annual rate of 5.25%. Use the formula in the problem above to find:
     i) the specific function giving the amount of money accrued in t years.
     ii) the amount of money accrued in 2 years.
     iii) the amount of money accrued in 10 years.

27. Some people reduce the size of their taxable estate before they die by giving money to friends or relatives. However, in some instances, a unified transfer tax T is assessed on these types of gifts. If the tax on any monetary gift x between \$40,000 and \$60,000 is computed according to the function $T(x) = 8200 + 0.24(x - 40{,}000)$, find $T(45{,}000)$ and $T(52{,}500)$ to the nearest cent.

28. The annual number (in millions) of new homes built is a function of the average interest rate $r\%$ on new home loans and is given by $h(r) = \dfrac{49 - 2r}{15}$. Find the number of new homes built when $r = 5, 8, 10$. Is this an increasing or decreasing function?

29. The cost, in dollars, of renting a car for a day is given by $C(m) = 0.3\,m + 10$, where m is the number of miles driven. What would it cost to drive the rental car 70 miles? 100 miles? 450 miles? Is the cost an increasing or decreasing function of the miles driven?

30. The resistance R, measured in ohms, of any aluminum wire to the flow of electric current is proportional to its length x and inversely proportional to its cross-sectional area a. Symbolically, this is written

$$R(x) = kx/a$$

For an aluminum ($k = 3.2 \times 10^{-8}$) wire that has a cross-sectional area of 10^{-6} square meter, what is the resistance of the wire (to the nearest tenth) if it has length 0.5 m? 50 m? 100 m? Is the resistance of a wire an increasing or decreasing function of its length? (A review of scientific notation is in the Appendix.)

31. The velocity v (in meters/sec) required for a satellite to remain in a circular orbit around the earth is a function only of the distance r (in meters) from the center of the earth given by

$$v(r) = \frac{2.0 \times 10^7}{\sqrt{r}}$$

Is this an increasing or decreasing function? Find the velocities (in m/sec) for $r = 9.0 \times 10^6$ m, 16.0×10^6 m, and 20.0×10^6 m.

32. The temperature B in degrees centigrade at which water boils is approximated by the function

$$B(h) = 100.86 - 0.04\sqrt{h + 431.03}$$

where h is the altitude (meters) above sea level. Find $B(1000)$ and $B(8500)$ to the nearest hundredth. Is this an increasing or decreasing function?

33. Several diseases are associated with resistance to the flow of blood to the fetal spleen in humans. Researchers have discovered that the index of splenic artery resistance can be modeled by

$$f(x) = 0.057x - 0.001x^2$$

where x is the number of weeks of gestation. What is the index of resistance after 8 weeks? After 30 weeks? [**Source**: Abuhamad, A. Z. et al., Doppler Flow Velocimetry of the Splenic Artery in the Human Fetus: Is It a Marker of Chronic Hypoxia?" American Journal of Obstetrics and Gynecology, Vol. 172, No. 3, March 1995, p. 820-825.]

34. If the brakes are applied to a car traveling at a velocity v, the distance d required to stop is related to the constant deceleration a on the car by

$$d = -v^2/2a.$$

Is d an increasing or decreasing function of a? Low tire pressure can lower the deceleration achieved by the car upon application of the brakes. If your tire pressure is lower than the advised standard, will the car travel a longer or shorter distance than it would with your tires at the correct pressure? [**Source**: National Highway Traffic Safety Administration, U.S. Department of Transportation]

35. The Greek astronomers in the time of Hipparchus assumed all stars to be the same distance from us (see Example 9) and so the magnitude numbers assigned to the stars were meant to be indicators of their sizes (see Figure 1.1.7). Dimmer (and therefore smaller) stars are classified by larger numbers. Suppose another ancient civilization instead adopted the hypothesis that all stars were the same size and therefore their differences in brightness in the night sky were the result of varying distances. If they also created a function from the stars to numbers by assigning larger magnitudes to dimmer stars, would a magnitude 2 star be closer or farther away than a magnitude 4 star? As magnitude increases under this scheme, does distance increase or decrease?

36. The elements in the *Periodic Table of the Elements* are grouped in rows and columns according to certain shared properties. One of these properties is *electronegativity* or the tendency of an element to acquire electrons in a chemical interaction. Electronegativity is a function of the atomic number that increases as atomic number advances along any row in the table (left to right) and decreases as atomic number advances along any column (top to bottom). The following list is a subset of the Periodic Table.

Li	Be	B	C	N	O	F
3	4	5	6	7	8	9

Na	Mg	Al	Si	P	S	Cl
11	12	13	14	15	16	17

K	Ca	Ga	Ge	As	Se	Br
19	20	31	32	33	34	35

According to the description of the function, does C (carbon) have more or less electronegativity than O (oxygen)? B (boron) have more or less than Ga (gallium)? Mg (magnesium) more or less than Cl (chlorine)? Rank the following elements from lowest to highest electronegativity:

Ca, As, F, N, P, K

37. The average fraction of the wind's energy that can be converted by a small wind turbine into electricity is given by $f(x) = 0.5x(2 - x)^2$ where x is the fraction by which the wind's speed is decreased as it strikes the turbine blades. Note that $0 \leq x \leq 1$. Find the fraction of the wind's energy converted for $x = 1/4$; $x = \frac{1}{2}$; $x = \frac{3}{4}$. [**Source:** U.S. Government Printing Office, *The Energy Book*]

38. In his classic *Moby Dick*, Herman Melville wrote about the use of the quadrant.

It was hard upon high noon; and Ahab, seated in the bows of his high-hoisted boat, was about taking his wonted daily observation of the sun to determine his latitude.

What other inputs might Ahab need in order to determine his latitude at sea as a function of the elevation of the noon sun?

However, Ahab grew weary of this equipment and settled for a much cruder formula.

"Curse thee, thou quadrant!" dashing it to the deck, "no longer will I guide my earthly way by thee; the level ship's compass, and the level dead-reckoning, by log and by line; these shall conduct me, and show me my place upon the sea."

What do you think is meant by "by log and by line" ?

1.2 Creation of Functional Expressions

We referred previously to the **action** of a function. By this we mean the steps taken for the function to process an element x in its domain in order to obtain the corresponding functional value $f(x)$. Suppose, for instance, we start with a number, say 2, and we are asked to perform some arithmetic operation or sequence of operations in order to transform it into 11. What might we do? A variety of options exist. For instance, we could simply add 9. Or, we could square 2 and then add 7. Or, we could multiply by 4 and add 3. All of these are acceptable procedures and, in fact , there exists a great many more.

Now, however, suppose we are asked to pick one of the above procedures that not only turns 2 into 11, but also 1 into 6, 3 into 16, 4 into 21, and 5 into 26. In other words, we need to identify the action of the correct function f that makes these assignments.

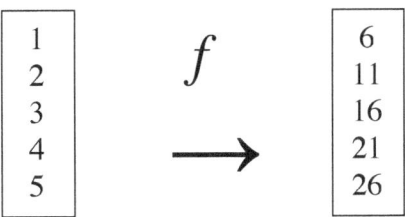

Figure 1.2.1 The assignments of a function f.

This type of problem is often faced by scientists, businessmen, engineers, and people from a variety of professions. Upon observing a list or table of paired values that are representative of an existing function, they create an algebraic expression for it. We have already considered this intuitively in the first section and now we look at it again a bit more formally.

Linear Functions

In the present case we must try to create a **formula** that represents the action of the function not just for an input of 2, but also for 3, 4, 5, and perhaps many other numbers. So this formula must encode the steps involved that will successfully produce the desired *outputs* of 11, 16, 21, 26, etc. The word "encode" implies use of a language, in this case the language of *algebra*. What must our formula do? It must pair each element in the domain of the function with the proper image. We need an independent *vari*able to "stand in" for a *vari*ety of inputs in contrast to a single unknown quantity. Using the symbol x, we let $f(x)$ stand for the corresponding image and construct a table (Figure 1.2.2) displaying the correspondences $f(1) = 6$, $f(2) = 11$, $f(3) = 16$, $f(4) = 21$, and $f(5) = 26$.

$$
\begin{array}{c|c}
x & f(x) \\
\hline
+1\left[\begin{array}{c}1\\2\end{array}\right. & 6\,]{+5} \\
+1\left[\begin{array}{c}2\\3\end{array}\right. & 11\,]{+5} \\
+1\left[\begin{array}{c}3\\4\end{array}\right. & 16\,]{+5} \\
+1\left[\begin{array}{c}4\\5\end{array}\right. & 21\,]{+5} \\
 & 26\,]{+5}
\end{array}
$$

Figure 1.2.2 A table of paired values for a linear function f.

Now, what do we do to x? Adding 9 to x works just fine if x will only be replaced by 2, but the fact that x will also be replaced by $3, 4, 5$, and so on spoils that process. Squaring and adding 7 doesn't work for all the inputs either. It is far more fruitful to examine how the functional value $f(x)$ changes as x changes. Each time x increases by 1, $f(x)$ increases by 5. The change in the functional value per unit change in the independent variable is called the **rate** (equal to the **slope** of the associated **graph** to be considered in the next section). In order to insure a constant rate of 5, we write

$$f(x) = 5x + k.$$

Now all we need is to determine the value of the constant term k by substituting any of the pairs from the table. For instance, $16 = f(3) = 5(3) + k$ means that $k = 1$. Therefore, the desired function is

$$f(x) = 5x + 1.$$

We can then check to see that this formula also correctly produces the other outputs:

$$f(1) = 5(1) + 1 = 6, \quad f(2) = 5(2) + 1 = 11, \quad f(4) = 5(4) + 1 = 21, \text{ and } f(5) = 5(5) + 1 = 26.$$

Notice how one input produces one output and *no more*. Recall how this is precisely the condition that defines a function. Functions provide definitive connections between sets. Why is this notion so important? If it is possible to construct a formula for the function, then knowing the correct function gives you the power to make an accurate *prediction* of a number in the range, based on being given a number in the domain. In our current example, the domain is $D = \{1, 2, 3, 4, 5\}$ and the range is $R = \{6, 11, 16, 21, 26\}$. Suppose instead that this is just a subset of a larger domain, say the interval $D = [1, 100] = \{x \mid x \text{ is a real number and } 1 \le x \le 100\}$? (See Appendix A for a complete description of intervals.) If we *assume that our rate remains constant* regardless of the input, then our new range is also an interval obtained by simply substituting these values for x in our function $f(x) = 5x + 1$ in order to get $R = [6, 501]$.

Any function with a constant rate is a member of an important class of functions known as **linear functions**. A linear function has the general form

$$f(x) = mx + k$$

where m is the rate and k is a fixed constant. If the rate is a positive number as in the current case, then $f(x)$ gets larger with increasing x and f is therefore an increasing function. On the other hand, if the rate is negative, then f is a decreasing function. A good example of a decreasing linear function concerns *depreciation* of machinery or other materials that are involved in running a business. If the value of a tractor on a farm decreases by $700 each year, the rate is -700 dollars/year and so we could write $V(t) = -700t + k$ using t (for time) as the independent variable. If the tractor was originally worth $42{,}000$, $V(0) = 42{,}000 = k$ and so the function would be $V(t) = -700t + 42{,}000$. Linear functions occur in a large variety of applications.

Example 1

Examine the following tables of paired values and create a formula for each to express the action of the functions that they represent. Assuming an expanded domain and constant rates, find $f(1.2)$, $g(20)$, and $h(45.6)$.

x	$f(x)$	x	$g(x)$	x	$h(x)$
−2	−2	−10	−67	1	4.4
−1	1	−5	−37	2	1.3
0	4	0	−7	3	−1.8
1	7	5	23	4	-4.9
2	10	10	53	5	−8.0

Solution

Recognition of the pattern is the key. As x increases by 1 in the first list $f(x)$ increases by 3. This implies that the rate of change of $f(x)$ is 3 and we can initially write $f(x) = 3x + k$. Then substituting, say 0, for x, we get that $k = f(0) = 4$ and so $f(x) = 3x + 4$.

In the second list $g(x)$ is increasing by a constant increment of 30 for each increase of 5 in x. Since the rate is defined as the change in the function per *unit* change in the independent variable, $m = \frac{30}{5} = 6$. We write $g(x) = 6x + k$ and then obtain k by substituting any of given pairs. Using $(0, -7)$, we get

$$g(0) = 6(0) + k = -7$$
$$k = -7$$

and so $g(x) = 6x - 7$.

In the last table, $h(x)$ is *decreasing* by 3.1 for each unit increase in x and so $m = -3.1$. Proceeding as before, $h(1) = -3.1 + k = 4.4$ and so $k = 7.5$ giving us $h(x) = -3.1x + 7.5$.

Finally, assuming the rates remain constant in expanded domains,

$$f(1.3) = 3(1.3) + 4 = 7.9 .$$
$$g(20) = 6(20) - 7 = 113 .$$
$$h(45.6) = -3.1(45.6) + 7.5 = -133.86. \quad \blacklozenge$$

Example 2

In a psychological study, one researcher recorded the following times which a sample of five adults needed to memorize a 9-digit sequence of numbers and letters according to age. The results are given in the following table.

Age(yr)	**Time** (sec)
38	29
42	31
46	33
50	35
54	37

The researcher would like to predict the memorization times for another older set of people before actually testing them. Based on the above sample, what would be the predictions for people of ages 56, 65, and 70?

Solution

An examination of these particular data reveals a *linear* connection between age a and memorization time T. The values of our independent variable here are not increasing by one at a time but rather by increments of 4. The rate of increase of time per year is

$$m = \frac{2 \text{ seconds}}{4 \text{ years}} = 0.5 \text{ sec/yr.}$$

We write $T(a) = 0.5a + k$ and find k by seeing that
$T(38) = 0.5(38) + k = 19 + k = 29$

implies that $k = 10$. We use our linear function $T(a) = 0.5a + 10$ to make the predictions.

$$T(56) = 0.5(56) + 10 = 38 \text{ sec.}$$
$$T(65) = 0.5(65) + 10 = 42.5 \text{ sec.}$$
$$T(70) = 0.5(70) + 10 = 45 \text{ sec.}$$

Of course, these values may not match the empirical results of the new survey very well, but they still provide an interesting source of comparative numbers for the researcher. ◆

The creation of a linear function does not have to arise from a table of values if the rate is available in some other way. In particular, if we *already* know that some function is linear, then we only need to know two data points in our table, that is, two sets of paired values.

Example 3

Acme Hardware manufacturing company has purchased a new drill press for $17,000. If it depreciates linearly and is worth $15,000 four years later, find a function that gives the value of the press at a time t years from now.

Solution

We only have two data points for this function, namely (0, 17000) and (4, 15000), but we have the advantage of *knowing* that the function is linear rather than having to determine that fact from some larger table. Hence, we find the constant rate by computing

$$m = \qquad\qquad = -500.$$

Now, since the value of the press is $V(t) = -500t + k$ and $V(0) = 17000$, we must have $k = 17000$ and so $V(t) = -500t + 17000$. ◆

Many types of functions do not change at a constant rate and are therefore known as *nonlinear* functions. However, they can still behave in a manner that allows us to create a predictive expression to represent them. There are numerous varieties of nonlinear functions but we shall limit ourselves to looking at just one major category of important examples, namely those functions in which the independent variable is an exponent.

Exponential Functions

We now return to the frequency table for musical pitches from the last section, but list the values differently in order to construct a common type of non-linear function.

Frequency Table

Octave	C	D	E	F	G	A	B
0	65.4	73.4	82.4	87.3	98.0	110.0	123.5
1	130.8	146.8	164.8	174.6	196.0	220.0	246.9
2	261.6	293.6	329.6	349.2	392.0	440.0	493.9
3	523.2	587.2	659.2	698.4	784.0	880.0	987.8
4	1046.4	1174.4	1318.4	1396.8	1568.0	1760.0	1975.6

Figure 1.2.3 Frequencies (Hz) of pitches in the chromatic scale for several octaves.

Example 4

Recall our table from Section 1.1 that identified each frequency by a pitch name. A second function can be produced from this chart by observing the well-known musical fact that, for any particular pitch, the next higher octave is obtained by doubling the frequency. For the pitch "A", if we arbitrarily identify octave 0 with 110 Hz, we have a function f such that

$$f(0) = 110, \quad f(1) = 220, \quad f(2) = 440, \quad f(3) = 880, \ldots \text{ etc.}$$

We can see a clear pattern here, namely that each frequency has 110 as a factor and a power of 2 as a factor.

$$f(0) = 110 \cdot 2^0 = 110$$
$$f(1) = 110 \cdot 2^1 = 220$$
$$f(2) = 110 \cdot 2^2 = 440$$
$$f(3) = 110 \cdot 2^3 = 880$$

As we shall see repeatedly, the recognition of patterns plays a major role in the observation and recording of the physical world around us. It is basically a three-part process. We use our five senses to observe a phenomenon, our intellect to sort and categorize the information, and mathematics to create a written record. In a very real sense, mathematics allows us to both extend and perfect our observations. In the current case, the domain of this function consists of the non-negative integers and the range consists of values that start at 110 and then increase by a *factor* of two each time the domain value increases by an additional increment of one. Such a function can be written down in a general form,

$$f(x) = 110 \cdot 2^x \ .$$

We can appreciate that this function "extends" our senses. Our ears have an upper limit for the frequencies they can hear, but our function tells us of the frequency associated with any octave of "A", no matter how high. ♦

The function demonstrated in the above example is known as an ***exponential function*** and is an accurate model of a wide variety of scientific and social phenomena. The general form is

$$f(x) = k\, b^x$$

where the coefficient k is any constant value and b is any positive number not equal to 1 referred to as the ***base***. An exponential function can be identified by the unique way its values increase. Consider the assignments in the table in Figure 1.2.4.

x	$f(x)$
-2	1/9
-1	1/3
0	1
1	3
2	9

+1 [-2 to -1] × 3
+1 [-1 to 0] × 3
+1 [0 to 1] × 3
+1 [1 to 2] × 3

Figure 1.2.4 A table of paired values for an exponential function.

For every unit increase in the independent variable here, the functional value increases *by a factor of* 3 rather than just an additional increment of 3 which is the case for a linear function. This implies that the base is 3 in this case and this characteristic is the hallmark of an exponential function. So we have initially that $f(x) = k \cdot 3^x$. As in the case of linear functions, we can then proceed to determine the value of k by substituting any of the given pairs of values. The easiest choice is to use $f(0)$ when it is available since $f(0) = kb^0 = k$. In the present case, $k = f(0) = 1$ and so our function must be $f(x) = 3^x$ which is consistent with the other pairings:

$$f(-2) = 3^{-2} = \frac{1}{9} \ ,$$
$$f(-1) = 3^{-1} = \frac{1}{3} \ ,$$
$$f(1) = 3^1 = 3 \ ,$$
$$f(2) = 3^2 = 9 \ .$$

Other values would be $f(-3) = 3^{-3} = \frac{1}{27}$, $f(3) = 3^3 = 27$, and so forth.

Example 5
Suppose our range values are slightly different than those above.

```
        x  | g(x)
+1 ⌈ -2  | 5/9  ⌉ x 3
+1 ⌈ -1  | 5/3  ⌉ x 3
+1 ⌈  0  | 5    ⌉ x 3
+1 ⌈  1  | 15   ⌉ x 3
+1 ⌈  2  | 45   ⌉ x 3
```

What has happened? As before we observe that the function here increases by a factor of 3 per unit increase in x. Therefore it is exponential with a base of 3, the same as above. Since each image under the previous function has been multiplied by 5, our new function must be $g(x) = 5 \cdot 3^x$. However, we don't need the first function handy in order to deduce the second one. Since we know $g(x) = k \cdot 3^x$, we can determine k from $5 = g(0) = k \cdot 3^0 = k \cdot 1 = k$. Alternatively, we could choose any other ordered pair to substitute into the function. For instance, since $g(2) = 45$, we get

$$45 = g(2)$$
$$= k \cdot 3^2$$
$$= k \cdot 9$$

which again gives us $\qquad k = 5 \qquad \blacklozenge$

Exponential functions appear in a wealth of diverse applications ranging from describing the decay of radioactive elements to the growth of money saved in a bank account. Originally most banks only added interest to a depositor's savings account once a year, known as annual compounding. (See the chapter on personal finance for more details.) Suppose you deposit $200 in a bank offering an annual interest rate of 5% and the interest is computed annually. At the end of one year, you would earn $(0.05)(\$200) = \$10.00$, which would increase your account to $210. Note that we could also obtain this figure by factoring.

$$200 + (0.05)(200) = 200(1 + 0.05) = 200(1.05) = 210.$$

If you then leave that amount on the bank for another year, this time you will earn $(0.05)(\$210) = \$10.50$, the extra $0.50 coming from the fact that now you are earning 5% on your first year's interest. This brings your total savings to $210 + 10.50 = \$220.50$. Again note that

$$210 + (0.05)(210) = 210(1 + 0.05) = 210(1.05) = \$220.50$$

which can be written

$$\mathbf{210}(1.05) = \mathbf{200(1.05)}(1.05) = 200(1.05)^2$$

where we have shown the replacement in bold-face. Observe the pattern that develops from continuing to leave your money in the same account for six years. Let $A(t)$ stand for the amount present at time t (in years).

At $t = 0$ years, $A(0) = \$200.00$

After $t = 1$ year, $A(1) = 200(1.05) = 210$
After $t = 2$ years, $A(2) = \mathbf{210}(1.05) = \mathbf{200(1.05)}(1.05) = 200(1.05)^2 = 220.50$
After $t = 3$ years, $A(3) = \mathbf{220.50}(1.05) = \mathbf{200(1.05)^2}(1.05) = 200(1.05)^3 = 231.53$
After $t = 4$ years, $A(4) = \mathbf{231.53}(1.05) = \mathbf{200(1.05)^3}(1.05) = 200(1.05)^4 = 243.10$

We see that we can easily form our own formula here since we see that the exponent for 1.05 in each case matches the number of years that have passed.

$$A(t) = 200(1.05)^t$$

In general, the amount $A(t)$ accumulated after t years from an initial principal P by compounding annually in an account earning interest at an annual rate r (as a decimal) is given by the following *exponential* function.

$$A(t) = P(1 + r)^t$$

This function needs only minor modification to conform to the modern practice of compounding more than once a year e.g. quarterly, monthly, daily, etc. For a fixed rate r, increasing the number of times per year that compounding is done causes the amount of money to grow more rapidly. (See Chapter 4 for a thorough treatment of the mathematics of finance.) As the number of compounding times per year gets infinitely large the expression for $A(t)$ becomes

$$A(t) = Pe^{rt}$$

where the number $e \approx 2.71828$ is a special number that occurs frequently in a wide variety of applications. Since e (named for the mathematician Leonhard Euler mentioned in the last section) occurs often in nature, e^{rt} is known as the **natural** exponential function with rate r. (Note that e^r serves as the base.) The function e^x is found on all modern calculators.

The growth of most organic populations from bacteria to rabbits to humans living in an environment of unlimited resources can also be modeled by the natural exponential function. If we think of new members of a population being produced almost continuously, then the above function provides a reasonable first estimation of how an initial ($t = 0$) population P increases if it has an annual growth rate r.

Example 6

In 1900, the population of the United States was 76 million people. Assuming natural exponential growth and a constant annual growth rate of 1.3%, estimate the population in the years 2010 and 2020.

Solution

The value of P always corresponds to an initial ($t = 0$) amount. If we let $t = 0$ correspond to the year 1900, then $P = 76$ (million) and $r = 0.013$ in the above function $A(t) = Pe^{rt}$. This gives us

$$A(t) = 76e^{0.013t}$$

in units of millions of people t years after 1900. Hence, our estimate for the year 2010 is

$$
\begin{aligned}
A(110) &= 76e^{0.013(110)} \\
&= 76e^{1.43} \\
&= 76(4.18) \\
&= 318 \qquad \text{(\textit{Rounding to three significant digits.})}
\end{aligned}
$$

or a population of 318 million people. Similarly, we estimate a population in the year 2020 of $A(120) = 76e^{0.013(120)} = 76e^{1.56} = 362$ million people. ◆

Our travels through this text will take us many places but we will have the same traveling companion throughout – that champion tool for expressing a definitive manner in which one variable entity depends expressly on another – the *function*. The ability to create a function connecting two sets of data is a valuable skill – one that allows us to describe, analyze, make predictions, and draw conclusions about phenomena that affect us and the world around us.

Exercise Set 1.2

Each of the following tables displays a sample of pairings under a function f. Identify each function as linear or exponential. In each case, construct a "formula" for f(x) under the action of the function.

1.

x	f(x)
1	11
2	12
3	13
4	14
5	15

2.

x	f(x)
1	2
2	4
3	6
4	8
5	10

3.

x	f(x)
1	9
2	11
3	13
4	15
5	17

4.

x	f(x)
1	6
2	12
3	18
4	24
5	30

5.

x	f(x)
1	5
2	11
3	17
4	23
5	29

6.

x	f(x)
1	−4
2	−8
3	−12
4	−16
5	−20

7.

x	f(x)
1	1
2	−3
3	−7
4	−11
5	−15

8.

x	f(x)
1	−3.5
2	2.5
3	8.5
4	14.5
5	20.5

9.

x	f(x)
-2	1.0
-1	4.5
0	8.0
1	11.5
2	15.0

10.

x	f(x)
-2	13.8
-1	12.9
0	12.0
1	11.1
2	10.2

11.

x	f(x)
-2	0.6
-1	-1.2
0	-3.0
1	-4.8
2	-6.6

12.

x	f(x)
0	2.8
2	9.2
4	15.6
6	22.0
8	28.4

13.

x	f(x)
0	−11.5
2	−1.3
4	8.9
6	19.1
8	29.3

14.

x	f(x)
-10	120.4
-5	60.4
0	0.4
5	−59.6
10	−119.4

15.

x	f(x)
-10	49.0
-5	25.5
0	2.0
5	−21.5
10	−45.0

16.

x	f(x)
-2	1/4
-1	1/2
0	1
1	2
2	4

17.

x	f(x)
-2	3/4
-1	3/2
0	3
1	6
2	12

18.

x	f(x)
-2	0.01
-1	0.1
0	1
1	10
2	100

19.

x	f(x)
-2	0.04
-1	0.4
0	4
1	40
2	400

20.

x	f(x)
-2	7/9
-1	7/3
0	7
1	21
2	63

21.

x	f(x)
-2	0.4
-1	2
0	10
1	50
2	250

22.

x	f(x)
-2	4
-1	2
0	1
1	0.5
2	0.25

23.

x	f(x)
-2	12
-1	6
0	3
1	1.5
2	0.75

24.

x	f(x)
-2	63
-1	21
0	7
1	7/3
2	7/9

25.

x	f(x)
-2	25,000
-1	5000
0	1000
1	200
2	40

26.

x	f(x)
-2	50,000
-1	5000
0	500
1	50
2	5

27. The value of a copying machine is $5200 when it is purchased. After 2 years, its value is $4750. Create a formula giving the value of the machine as a linear function of time and use it to find the value after 5 years and 8 years.

28. You hail a taxi and inquire about the fare. The taxi driver informs you that your ride will cost a flat fee of $2.00 plus $1.25 per minute. Create a formula that gives you the cost of the ride as a function of the length of time spent in the car. What is the cost of a 28-minute ride?

29. The pipes in your laundry room are behaving mysteriously and you need to hire someone to fix them. The local plumber charges a minimum of $25 for any visit plus $32.50 per hour. Create a formula that gives you the cost as a function of the number of hours spent fixing your plumbing. How much will it cost if the plumber needs 4.5 hours to fix your pipes?

30. The numbers of pairs of birds (all species) that nested in the Franklin Bay rookery over four years are given according to the following table. Create a formula that gives the number of nesting pairs as a function of years after 2006. At the current rate how many pairs will be nesting there in 2012? 2015? By what year will no birds be nesting in this rookery?

Year	Nesting Pairs
2006	830
2007	752
2008	674

31. In one European country the federal income taxes for working residents are given below for some income brackets.

Annual Income ($)	Tax ($)
25,000 up to 30,000	2500 + 25% of excess over 25,000
30,000 up to 35,000	3750 + 28% of excess over 30,000
35,000 up to 40,000	5150 + 32% of excess over 35,000
40,000 up to 45,000	6750 + 36% of excess over 40,000

Find the expression which gives the tax as a function of annual income for the 30,000 to 35,000 bracket and use it to compute the tax owed by someone who earned $32,450.

32. Use the table in the above exercise to find the expression which gives tax as a function of annual income for the 35,000 to 40,000 bracket and use it to compute the tax owed by someone who earned $37,250. What is the tax owed by someone who earned $41,900? $44,730?

33. This table shows approximations to the life expectancy of a 25-year-old male smoker based on his daily cigarette consumption. Find a linear expression that estimates the life expectancy $f(x)$ as a function of the number of cigarettes smoked per day. Interpret the meaning of the rate.

Daily Cigarettes Smoked	Life expectancy (years)
0	73.7
10	71.6
20	69.5
30	67.4
40	65.3

34. Use your function from the previous problem to estimate the life expectancy of a 25-year-old male who smokes 25 cigarettes per day; 32 cigarettes per day.

35. If x is the temperature in degrees Celsius and $F(x)$ is the corresponding temperature in degrees Farenheit, then we know that $F(0) = 32$ and $F(100) = 212$. Find the formula which gives the action of this linear function and use it to find $F(45)$.

36. Angelita owns and manages a print shop. She specializes in printing brochures and charges customers a flat set-up fee of $45 plus a cost per brochure according to the quality of paper used. The table below lists total cost per type for quantities of 50 and 100. A customer requests 2000 brochures printed on the best possible paper without exceeding $450. Find the three linear functions that represent the cost of using each type of paper. Which type of paper should Angelita use?

Quantity	Type A	Type B	Type C
50	$52.50	$55.00	$57.50
100	$60.00	$65.00	$70.00

37. Suppose you deposit $750 in a savings account at a bank offering an annual interest rate of 4.2% compounded annually. Form the expression that gives the amount of money accrued as a function of time t in years. What amount of money will you have in that account in 3 years? 5 years?

38. George deposits $2800 with an investment firm that guarantees an annual return of 6.5%. If we assume annual compounding, form the exponential function showing the amount of money George will have in t years. How much will he have accumulated in 4 years? 7 years?

39. To attract business, *Joel's Bank* is advertising an annual interest rate of 10% for your first year. After that, the rate drops to 3.5%. If the compounding is done annually, compute the amount accrued in 6 years from an initial deposit of $2000.

40. *Maura's Bank* (Joel's competitor in the above problem) does not have a higher one-year interest rate, but guarantees a 4.5% annual rate every year. How much money would you accumulate from a deposit of $2000 in 6 years at this bank? In which bank would you deposit your money? What if you were going to leave it there for 7 years?

41. In 2012 the country of Herzon had a population of 1.85 million and was growing according to a *natural* exponential function with an annual growth rate of 2.3%. If that growth rate remains constant, what will be the population of Herzon in the year 2014? 2016? 2018?

42. In 2007 the county seat of Abington Township had a population of 1035 and was growing according to a natural exponential function with an annual growth rate of 0.4%. If that growth rate remains constant, what will be the population in the year 2010? 2017? 2030?

43. Statistics produced by the United Nations forecast an annual approximate world population growth rate. Assuming an annual rate of 1.2% and a world population of 7.0 billion in 2011, estimate the world population in 2015 and 2020. If this annual rate then drops to 1.1%, predict the world population in 2025. [*Source*: World Population Data Sheet (2011). Washington DC: Population Reference Bureau, Inc.]

44. The population of the country of Uganda in 2011 was 34.5 million people and its growth rate was 3.4%. [*Source*: World Population Data Sheet (2011). Washington DC: Population Reference Bureau, Inc.] Estimate the population in 2015 and 2020.

45. The population of the country of Mexico in 2011 was 114.8 million people and its growth rate was 1.4%. [*Source*: World Population Data Sheet (2011). Washington DC: Population Reference Bureau, Inc.] Estimate the population in 2015 and 2020.

46. The population of the country of Finland in 2011 was 5.4 million people and its growth rate was 0.2%. [*Source*: World Population Data Sheet (2011). Washington DC: Population Reference Bureau, Inc.] Estimate the population in 2015 and 2020.

47. The ***doubling time*** of a population is the amount of time it requires for that population to double in size. The ***Rule of 70*** states that if the annual growth rate is r percent, then the doubling time is approximately equal to $\dfrac{70}{r}$ (where r is left as a percentage and not converted to a decimal). Estimate the doubling times for Uganda, Mexico, and Finland using the rates from the previous problems.

48. If the doubling time for a population is 35 years, what is its annual growth rate? Use the rate you compute in the natural exponential function to verify that a population P grows to $2P$ in 35 years.

49. Certain *decreasing* populations can be modeled by exponential functions having a base that is less than one. Suppose 512 cupcakes were placed on a table in the cafeteria of Marywood University at 8:00 am. If only half of the cupcakes present at the beginning of each hour are still remaining at the end of the hour, how many were left by 5:00 pm? Find an expression that gives the number of cupcakes as an exponential function of the number of hours past 8:00 am.

50. Here are the population figures for three cities over the same time period. Two of these cities grew linearly and one grew exponentially. Create the expression giving the population for each city as a function of time letting $t = 0$ correspond to 2005.

Year	Sometown	Anytown	Boomtown
2005	10,500	37,000	40,000
2006	11,200	34,700	60,000
2007	11,900	32,400	90,000
2008	12,600	30,100	135,000
2009	13,300	27,800	202,500

51. Suppose *Smooth as Silk* paint manufacturer buys a new large mixing vat for $22,000. This vat depreciates non-linearly because its value at the end of each year is 90% of what it was worth at the start of the year. Find the decreasing exponential function that gives the value of the vat V in terms of time t in years after the initial purchase. How much will the vat be worth in 12 years?

52. If the temperature is constant, then the atmospheric pressure p (lb/in^2) depends on the altitude x (ft) above sea level according to $p = 15\,(0.8)^{(x/5000)}$. Is this a decreasing function? What is the atmospheric pressure at sea level ($x = 0$)? What is the pressure at 10,000 ft?

53. Radioactive substances decay over time and so the amount A at any time t is a decreasing function of t. Given an initial amount P, the amount A remaining at time t can given by the natural exponential function that has a negative rate r. For Plutonium 241, the function is given by

$$A(t) = Pe^{-0.053t}.$$

Starting with an initial amount of 20 gm of Plutonium 241, how much will remain after 5 years? After 8 years?

54. In a city with population P, the number of people N who have heard about a news bulletin broadcast over radio and television after t hours is $N(t) = P(1 - e^{-0.3t})$. In city of 50,000 residents, how many people have heard of a major earthquake 5 hours after it was broadcast? 8 hours?

55. Studies have shown that income is affected by the amount of mathematics courses an individual has taken. For example, if x represents the number of years of calculus classes taken, then the salary S earned by that person might by given by $S(t) = 45{,}000e^{0.195x}$ dollars. What is the salary of someone who has had 1.5 years of calculus? [**Source:** *Review of Economics and Statistics* 86]

It is useful to be able to formulate the action of a function from a verbal description. For instance, $f(x) = (2x + 7)^2$ is the function described by the procedure:
>   *1. Pick any number.*
>   *2. Double it and then add 7.*
>   *3. Square the result.*

Formulate the action of the function described by each of the following procedures.

56.　　1. Pick any number.
　　　　2. Square it and then multiply by 13.
　　　　3. Add 20 and then divide the result by 9.

57.　　1. Pick any number.
　　　　2. Subtract 4 and then square it.
　　　　3. Multiply the result by 5.

58.　　1. Pick any number.
　　　　2. Add 45 and then raise it to the fourth power.
　　　　3. Multiply the result by 28.

59.　　1. Pick any number (greater than or equal to –3).
　　　　2. Add 3 and then divide by 8.
　　　　3. Take the square root of the result.

60.　　1. Pick any number and square it.
　　　　2. Multiply the result by 5 and then add 3 times the original number you picked.
　　　　3. Add 12 to the result.

61.　　1. Pick any number (except 1) and add 6.
　　　　2. Raise the result to the fourth power and add 7.
　　　　3. Take the square root of the result.
　　　　4. Divide that result by the difference of the original number and 1.

1.3 The Graph of a Function

Now that we have had some experience forming expressions for functions from raw data, our next step is to acquire skill at analyzing the formula we have created. The whole point in knowing a formula is that it assists in knowing how the function behaves. We want to be able to answer questions like: Where does it increase? Where does it decrease? Does it achieve a maximum value and, if so, where? Does it achieve a minimum value and, if so, where? Does it ever become zero? Your symbolic expression is a description of a real physical process and answers to questions like these form the basis for a thorough understanding of that process. In fact, finding these answers is one of the main reasons why a person formulates a function in the first place.

That brings us to the role of a *graph* of a function. As you learned in algebra, the graph of an equation in two variables consists of ordered pairs of numbers satisfying the equation and is usually pictured as a set of points existing in a two-dimensional coordinate system. The use of coordinate systems is credited in part to the great French mathematician and philosopher **Rene Descartes** (1596–1650) due to his classic work *La Geometrie* which brought the force of algebra to bear on problems in geometry.

Descartes had the good fortune, in an era filled with poverty and suffering, to be born to an enlightened, caring, and wealthy father. The combination led to a catering of Descartes' early frail health and this bright, curious boy was allowed to spend the majority of the mornings of his youth at home. These times of solitude fostered a view of the world rooted in rational skepticism. Hence, along with such famous contemporaries as Galileo and Pascal, he

www.shutterstock.com · 24316807

was a pioneer in accruing knowledge strictly through experimentation and the application of mathematical reasoning. As a young man with his health restored, he split his time, amazingly, between contemplating science and mathematics and being a mercenary soldier for various kings and despots across the breadth of Europe. He doubted most of the standard axioms of knowledge of his day and assembled his own view of the universe brick by logical brick.

Although Descartes never fully employed the coordinate system which bears his name, it is said that the wondrous idea of representing geometric figures by equations came to him on a Bavarian battlefield in a dream of almost mystic potency on November 10, 1619. This date is now often referred to as the birthday of analytic geometry. By associating equations with curves, Descartes and others, notably the great **Pierre de Fermat** (1601–1665), used the new "coordinate geometry" as a tool to solve geometric construction problems which were considered essential to understanding some basic scientific questions. As it turned out, this "tool" had a tremendous impact on how mathematics came to be used to describe nature, one that is still being felt today. For that reason, the standard *xy*-plane of algebra is often referred to as the *Cartesian* coordinate system or plane.

The graph of a function gives a visual representation of how the functional values in the range respond to changes in values in the domain. If x is the independent variable for a function f and y is the dependent variable, then we say that "y is a function of x" and we define the **graph** of f to be the set $\{(x, y) \mid x \in \text{domain of } f \text{ and } y = f(x)\}$. In other words, the graph is the set of ordered pairs of values where the pairing is done according to the functional assignments.

Let's pause and make certain of the meaning of this definition. If $(2, 9)$ is a point on the graph of the function f, then that means precisely that $f(2) = 9$. If $(3.2, 51.6)$ is on its graph, then $f(3.2) = 51.6$. If $f(-5.8) = 101$, then $(-5.8, 101)$ must be on the graph. This one concept provides the framework for representing a functional relationship between two sets of numbers as a picture in the Cartesian plane.

Example 1

Graph the linear function $f(x) = 3x - 4$.

Solution

From our definition we see our graph consists of ordered pairs (x, y) such that $y = 3x - 4$. The easiest approach is to set up a table with a few arbitrary values of x. Suppose we let x equal each of the integers running from, say, -2 to 2.

x	$f(x)$
-2	-10
-1	-7
0	-4
1	-1
2	2

Now we plot these points in the Cartesian plane. After plotting the five points from our table, we extend the domain to all real numbers by connecting those points with a smooth curve. In this case, a line results since our function is linear. An analysis of the above table reveals that the corresponding constant rate of change is 3. It is no coincidence that 3 is also the coefficient of the x-term, which you may recall from algebra is known as the **slope** of the graph of the function. So the rate of a linear function and the slope of its graph are always the same value. Since this implies that the slope m is the rate of increase of the functional value per unit increase of the independent variable, it can always be computed from *any* two points (x_0, y_0) and (x_1, y_1) on the graph.

$$m = \frac{\text{vertical change}}{\text{horizontal change}} = \frac{\text{change in } y\text{-coordinates}}{\text{change in } x\text{-coordinates}} = \frac{y_1 - y_0}{x_1 - x_0}$$

In the present case, if we choose, say $(2, 2)$ and $(-1, -7)$, we compute the slope

$$m = \frac{-7 - 2}{-1 - 2} = \frac{-9}{-3} = 3.$$

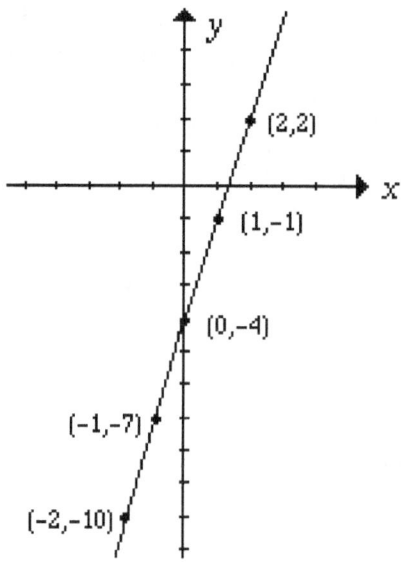

Figure 1.3.2 Graph of $f(x) = 3x - 4$.

We recognize this function as an increasing function, which in turn corresponds to the fact that it has a positive slope. If a linear function possesses a negative slope, it must necessarily be a decreasing function since a negative rate implies that the functional values are dropping as the independent values increase.

The graph of an exponential function looks very different.

Example 2

     Graph the exponential function $g(x) = 3 \cdot 2^x$.

Solution

     Again we set up a table with integer values of x running from -2 to 2.

x	$g(x)$
-2	3/4
-1	3/2
0	3
1	6
2	12

We plot these points on the xy-plane and draw a smooth curve.

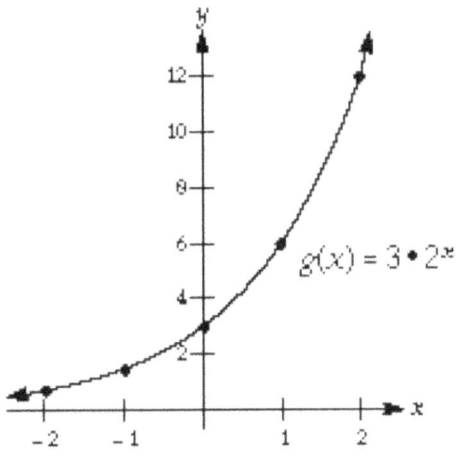

Figure 1.3.3 Graph of $g(x) = 3 \cdot 2^x$

As x increases by an increment of 1, the linear function in Example 1 increases by an *increment* of 3 (the slope), while the exponential function in Example 2 increases by a *factor* of 2 (the base). As explained previously, this property distinguishes linear from exponential functions, and Figures 1.3.2 and 1.3.3 illustrate how it affects the shapes of the corresponding graphs. Notice, in particular, that the graph lies entirely above the x-axis since an exponential function $f(x) = k\, b^x$ ($k > 0, b > 0$) is always positive. More specific features that distinguish the graph of every exponential function will be discussed in more detail in the next section.

Example 3

Every function of the form $f(x) = ax^2 + bx + c$ (where $a \neq 0$) is called a *quadratic* function. Graph the quadratic function $f(x) = x^2 - 4$. (In this case, $a = 1$, $b = 0$, and $c = -4$.)

Solution

We form a table of values with x running through the integers from –3 to 3.

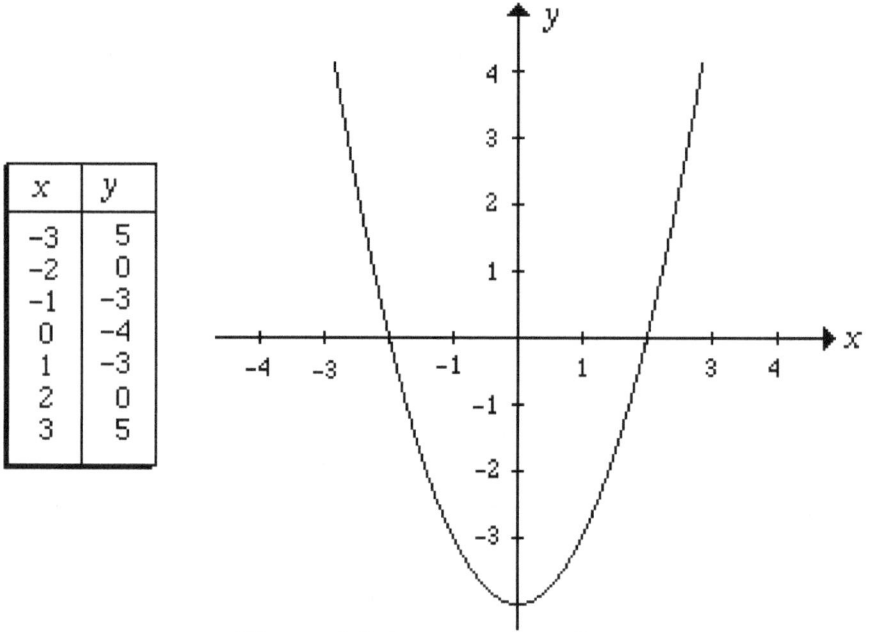

x	y
-3	5
-2	0
-1	-3
0	-4
1	-3
2	0
3	5

Figure 1.3.4 Graph of $f(x) = x^2 - 4$

The graph of every quadratic function is a curve known as a ***parabola***. Note that this parabola crosses the x-axis at $x = -2$ and 2 and crosses the y-axis at $y = -4$. These are known as the *intercepts* of the graph. The graphs of quadratic functions will be studied in greater detail in the next section. ◆

Functions can also be graphed with a domain that has been restricted. In this case, we simply do not include any points on the graph whose x-coordinate lies outside the given domain.

Example 4

Graph the function $g(x) = -\tfrac{1}{2} x + 5$ over the closed interval domain $[-2, 6]$.

Solution

This is a linear function with slope $-\tfrac{1}{2}$ and crossing the y-axis at 5. Note that no points are included whose x-coordinate is less than -2 or greater than 6.

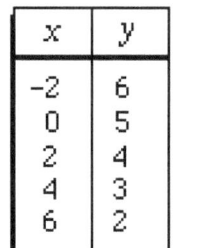

x	y
-2	6
0	5
2	4
4	3
6	2

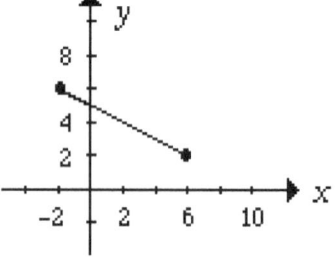

Figure 1.3.5 Graph of $g(x) = -(1/2)x + 5$ over the restricted domain $[-2, 6]$.

48

The graph of any function presents us with a picture of its behavior. The key is to remember that the y-coordinate of any point on the graph is the functional value of the x-coordinate. For example, by examining the graph in Figure 1.3.6, we can conclude that $f(1.5) < f(2)$ simply because the y-coordinate of the point on the graph with x-coordinate 1.5 is less than the y-coordinate of the point with x-coordinate 2. There are literally an infinite number of observations we could make e.g. $f(3.2) > f(3.7), f(2.5) > f(4.8)$, or even that $f(3)$ appears to be the largest value the function attains. Such a value is called the **maximum** value of the function and will be discussed further in Section 1.3.

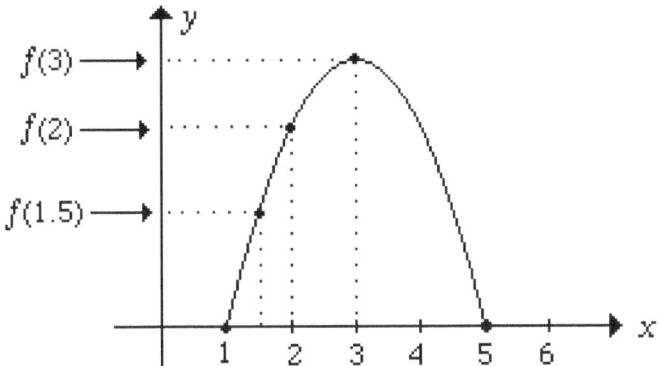

Figure 1.3.6 From this graph we see that $f(1.5) < f(2)$ and that $f(3)$ is the maximum value of the function.

Be aware that not every curve in the Cartesian plane is the graph of a function. Consider Figure 1.3.7. Is this the graph of a function?

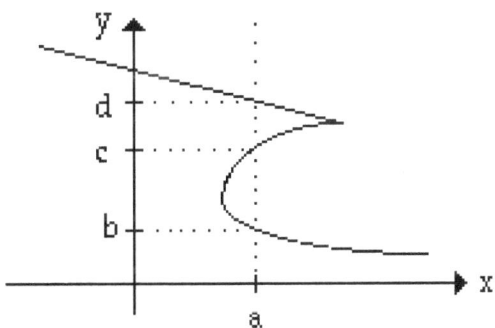

Figure 1.3.7 Vertical line test.

If the above curve did represent a function, say $g(x)$, would $g(a)$ be equal to $b, c,$ or d? There is no way to choose. *The* requirement for a relation to be a function is that the functional pairing with each value of the independent variable be unambiguous. In this case, there is not a unique image which we can assign to a. Pictorially, we see that the vertical line above a intersects the curve more than once and so we have the condition which is the basis for a handy check known as the **vertical line test**.

A curve in the Cartesian plane is the graph of a function if and only if no vertical line intersects the curve more than once.

We make one final observation about finding the domain and range of any function. Consider the graph of the linear function $f(x)$ in Figure 1.3.8. To determine the domain we must ask, "What are the *x*-coordinates of the points on the graph?" Imagine a flashlight being aimed at the graph from the top (and the bottom if it crosses the *x*-axis). The shadow projected on the *x*-axis is precisely that set of *x*-coordinates and therefore constitutes the domain. From the figure, convince

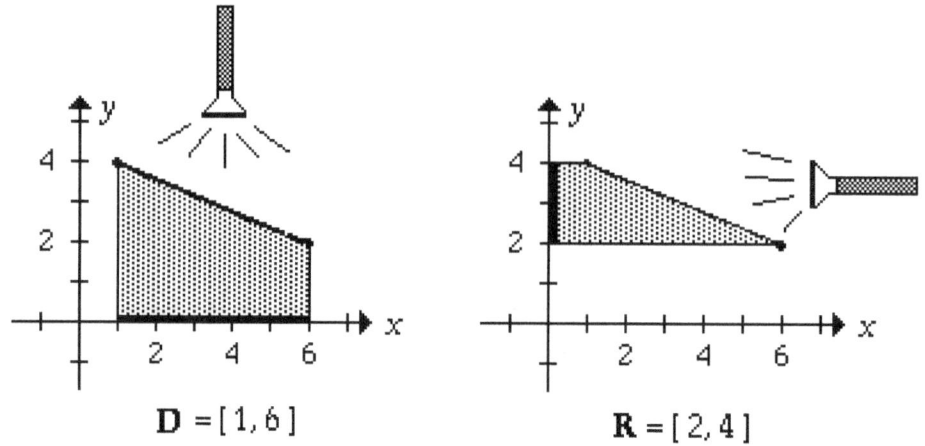

Figure 1.3.8 Determining the domain and range of a function.

yourself that the domain of *f*(*x*) is the interval [1, 6]. Similarly, we determine the range by asking, "What are the *y*-coordinates of the points on the graph?" By shining our flashlight from the right (and the left if the graph crosses the *y*-axis) this time, we can determine the range by looking at the curve's shadow on the *y*-axis. In the figure we see that the range of *f*(*x*) is the interval [2, 4]. In the next section we will learn more how the graph of a function reveals a great deal of information about its behavior.

Example 5
    Find the domain and range of the function whose graph is given in Figure 1.3.9.

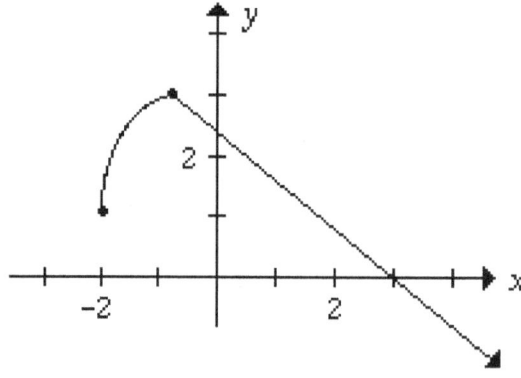

Figure 1.3.9

Solution
    The arrow in the above graph indicates that the line continues forever in that direction. Consequently the imaginary shadows produced by our flashlight would yield the interval [−2, ∞) as the domain and (−∞, 3] as the range of this particular function. ♦

Graphing Calculators

Calculator technology has greatly expanded our abilities to draw and use graphs in order to help us determine the behavior of functions. Graphs that are difficult and time-consuming to create by hand can be done in minutes on a calculator with a graphing capability. For instance, the five buttons beneath the window of a typical Texas Instruments calculator are shown below.

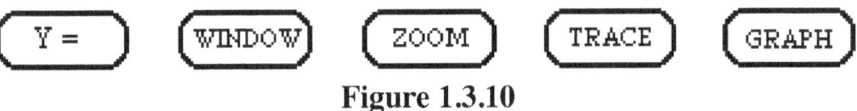

Figure 1.3.10

The first button allows you to input any function that can be composed from the functions available on the calculator. The second button allows scaling of the axes that will appear in the window. When you have assigned the necessary commands with these two features, pressing the **GRAPH** button produces the graph of your function. After the curve appears, you can check the coordinates of any of the points on it by pressing **TRACE** and moving the flashing point with the cursor arrows. Finally, you may zoom in and out to view the graph with different levels of magnification with **ZOOM**.

The speed of graphing with a calculator allows us to more easily learn about the effects on any graph brought about by changing the value of a coefficient, power, or other parameter.

Example 7

Use a calculator to graph $y_1 = 0.25\,x^2$, $y_2 = 0.5x^2$, $y_3 = x^2$, and $y_4 = 3x^2$ all in the same window. Adjust your axes so that **Xmin** = –5, **Xmin** = 5, **Xsc1** = 1, **Ymin** = –1, **Ymax** = 8, and **Ysc1** = 1. (**Xscl** and **Yscl** set the distance between tickmarks on the axes.) What do you conclude about the effect on the graph of $y = ax^2$ as a increases?

Solution

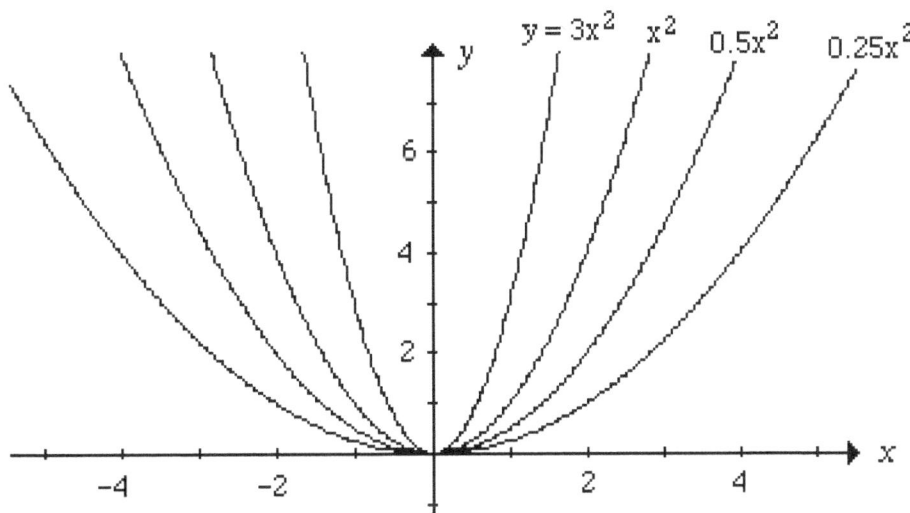

Figure 1.3.11 Graphing on a calculator.

The width of the parabola decreases as the coefficient a increases. ◆

51

Exercise Set 1.3

Graph each of the following functions.

1. $f(x) = -2x + 3$

2. $g(x) =$

3. $h(x) = \quad x + 5$

4. $y(x) = x^2 + 2$

5. $A(x) = 2x^2 - 5$

6. $P(x) = \quad x^2 - 3$

7. $f(t) = -t^2 + 9$

8. $g(x) =$

9. $f(x) = 3^x$

10. $w(t) = 6$

11. $r(t) = 2\,(0.8)^t$

12. $F(x) = 3 \cdot 2^x$

Graph each of the following functions over the given restricted domain.

13. $k(x) = -3x + 1$; $D = [-3, 2]$

14. $g(x) = \quad x - 5$; $D = (-\infty, 8]$

15. $r(t) = -t^2 + 9$; $D = [-4, \infty)$

16. $f(x) = 2x^2 - 6$; $D = [-1, 2]$

17. $h(t) = 5 \quad$; $D = [-2, 3]$

18. $y(t) = 4\,(0.3)^t$; $D = [-3, 6]$

In the next four problems determine whether the accompanying statements concerning the function of the corresponding graph are true or false.

19.   i) $f(-3) < f(-2)$ ii) $f(1) < f(3)$ iii) $2 < f(0) < 3$

     iv) $f(3) > 0$ v) $f(-1)$ is the maximum value of the function

     vi) $f(2) = 0$ vii) $f(0.5) < f(2.5)$ viii) $f(4) < 0$

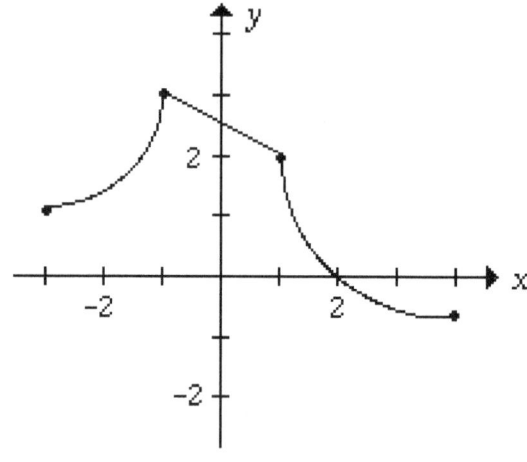

20.   i) $f(-1) < f(1)$ ii) $f(1) < f(2)$ iii) $f(2) < f(3)$ iv) $f(0) > 0$

       v) $f(-2) < 0 < f(-1)$ vi) $f(-1.9) > f(1.9)$ vii) $f(2)$ is the maximum value of f.

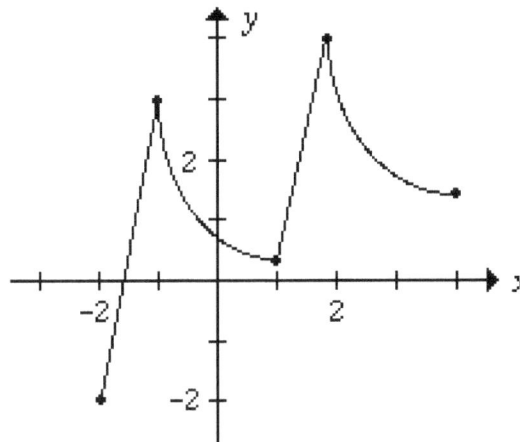

21.   i) $f(-2) < 0$ ii) $f(0) < f(1)$ iii) $f(0) > 0$ iv) $f(2) < f(3)$

       v) $f(0)$ is the minimum value of f.   vi) $f(-2)$ is the maximum value of f.

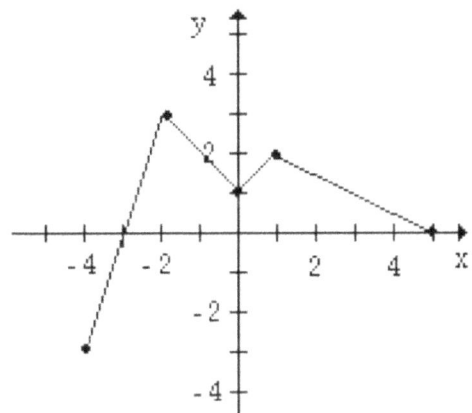

22.   i) $f(x) = f(-3)$ for $x < -3$.   ii) $f(-3) < f(-2)$ iii) $f(2) > f(3)$ iv) $f(3) < f(4)$

       v) $f(x)$ has no maximum value.   vi) $f(x) < 0$ for $x > 0$.

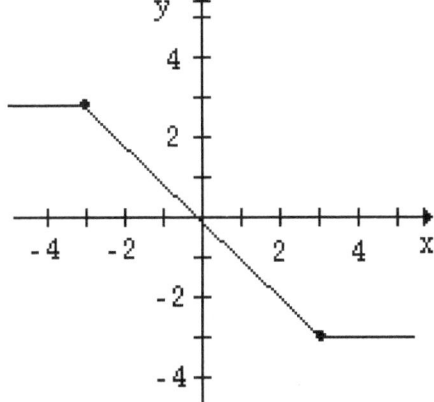

For each of the following graphs determine whether or not it represents a function and, if so, find the domain and range.

23.

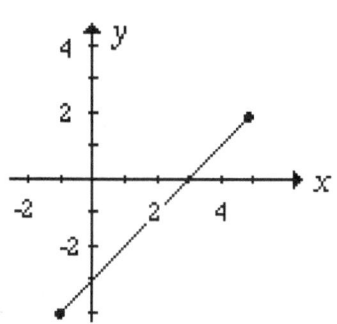

24.

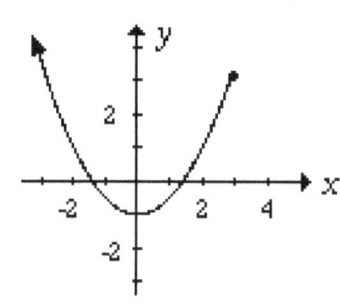

25.

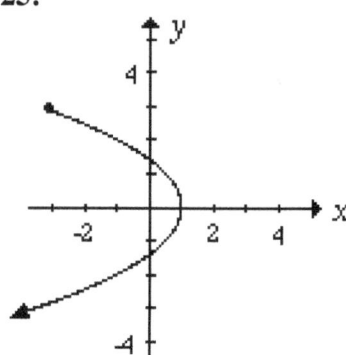

26.

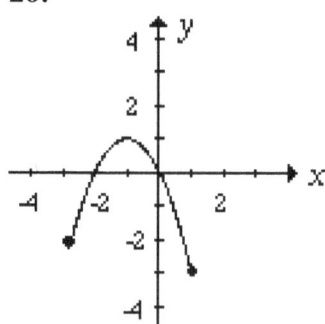

27.

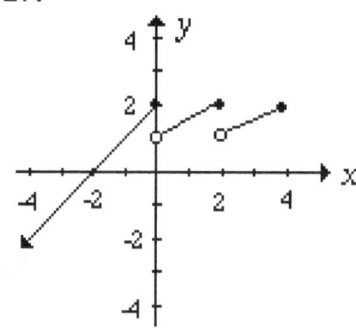

28.

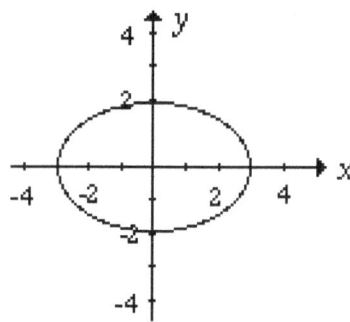

29.

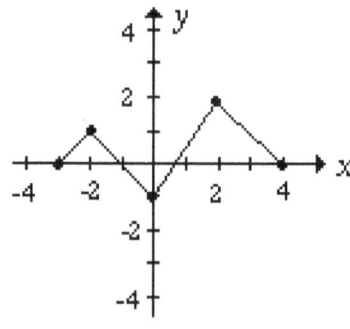

30.

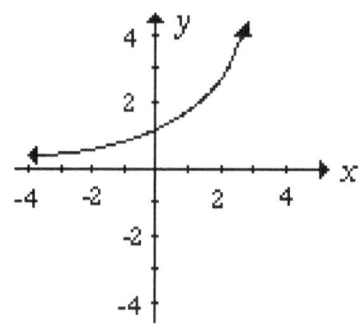

31.

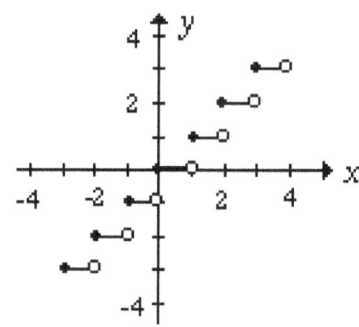

32.

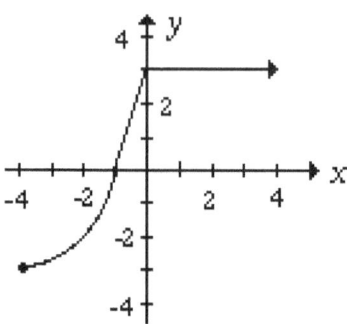

33.

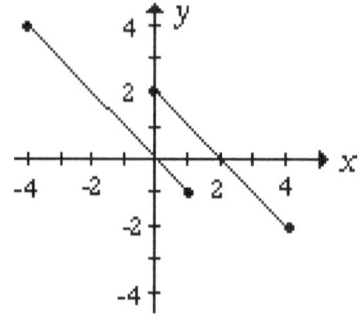

34.

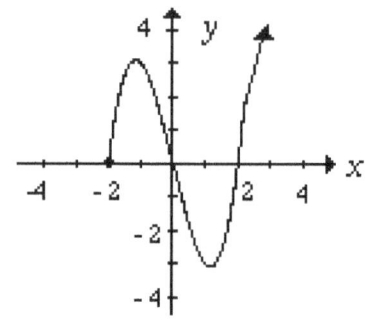

Sketch the graph of each of the following functions by reproducing the graph obtained using a **calculator**. *Graph each of the three functions for each problem in the same window sized according to the directions. The accompanying questions are concerned with associating features of the graph with values of parameters in the function.*

Linear functions

Size your window so that both axes run from –4 to 4 and the scale is set to 1.

35. $f(x) = 2x + 1$ $\qquad\qquad$ $g(x) = 2x + 3$ $\qquad\qquad$ $h(x) = -0.5x + 2$

i) Which two of these graphs are parallel? What do those functions have in common?

ii) Which graph has a y-intercept of 2?

iii) Which of the functions is decreasing?

36. $f(x) = 3x + 3.5$ $\qquad\qquad$ $g(x) = -2x - 4$ $\qquad\qquad$ $h(x) = 3x + 1$

i) Which two of these graphs are parallel? What do those functions have in common?

ii) Which graph has an x-intercept of –2?

iii) Which of the functions is decreasing? What parameter indicates that?

Exponential Functions

Size your window so that the x-axis runs from –4 to 20 and the y-axis runs from –1 to 10. Set both scales to 4.

37. $f(x) = 2 \cdot (1.1)^x$ $\qquad\qquad$ $g(x) = 2 \cdot (1.5)^x$ $\qquad\qquad$ $h(x) = 1.5 \cdot 2^x$

i) Which of these functions has a y-intercept of 2?

ii) In the exponential function $f(x) = k\, b^x$, which parameter will be equal to the y-intercept of the graph?

iii) Explain why the graph of an exponential function has no x-intercepts.

38. $f(x) = (0.25)^x$ $\qquad\qquad$ $g(x) = 2 \cdot (0.25)^x$ $\qquad\qquad$ $h(x) = 3 \cdot (0.5)^x$

i) Are these three functions increasing or decreasing?

ii) By comparing the bases of these functions with those in the previous exercise, formulate a rule for the base that determines whether a function is increasing or decreasing.

iii) What number does each function approach as x becomes infinitely large?

Natural Exponential Functions

Size your window so that the x-axis runs from –1 to 40 (scale = 10) and the y-axis runs from –1 to 15 (scale = 5).

39. Graph the natural exponential function $f(t) = 5\, e^{rt}$ both for $r = 2\%$ and then for $r = 3\%$.

i) If these functions represent the population growth of two towns (in thousands) and t is measured in years, what is the initial ($t = 0$) population?

ii) Which of these functions increases more rapidly?

iii) How many years would it take for $f(t)$ to reach 10 for each of the given values for r?

40. Graph the natural exponential function $f(t) = 7.5 \, e^{.045t}$. Use the graph to estimate the values of t for which $f(t) = 5, 10$, and 15.

Quadratic functions
Size your window so that both axes run from –4 to 4 and the scale is set to 1.

41. $f(x) = x^2 - 3$ $\qquad\qquad$ $g(x) = x^2 + 2$ $\qquad\qquad\qquad$ $h(x) = -x^2 + 4$
i) What main feature of the third parabola differs from the first two?
ii) What is different in the expression for the third function that causes this difference?
iii) What is the y-intercept of each graph?

42. $f(x) = x^2 + 2x - 3$ $\qquad\qquad$ $g(x) = -0.75x^2 + x + 4$ $\qquad\qquad$ $h(x) = 2x^2 - 3x - 1.5$
i) Estimate the vertex of each parabola using the TRACE button on your calculator.
ii) If the coefficient of the x^2 term is positive, does the graph open up or down? Negative?
iii) Is the y-intercept of the graph of a quadratic function $f(x) = ax^2 + bx + c$ equal to a, b, or c ?

43. Most graphing calculators can draw more than one graph on the same screen. Graph the following linear functions on the same set of axes:

$$y_1 = -2.6x \qquad y_2 = -0.5x \qquad y_3 = 0.3x \qquad y_4 = x \qquad y_5 = 2x$$

What happens to the graph as the slope of the function increases?

44. Graph the following exponential functions on the same set of axes. Size your window so that the x-axis runs from –5 to 5 and the y-axis runs from –1 to 8.

$$y_1 = 1.2^x \qquad y_2 = 1.4^x \qquad y_3 = 1.7^x \qquad y_4 = 2^x$$

What happens to the shape of the graph as the value of the base increases? How does this correspond to the behavior of the function?

Find a formula (by experimentation) for each function whose graph is given below.

45.

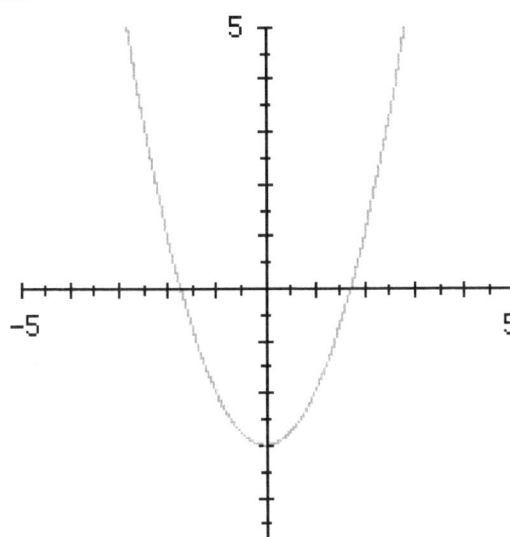

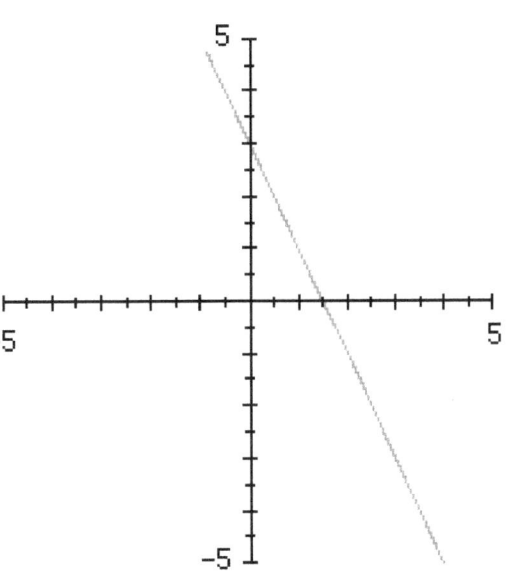

i) $f(x) =$ _____

ii) $f(x) =$ _____

46.

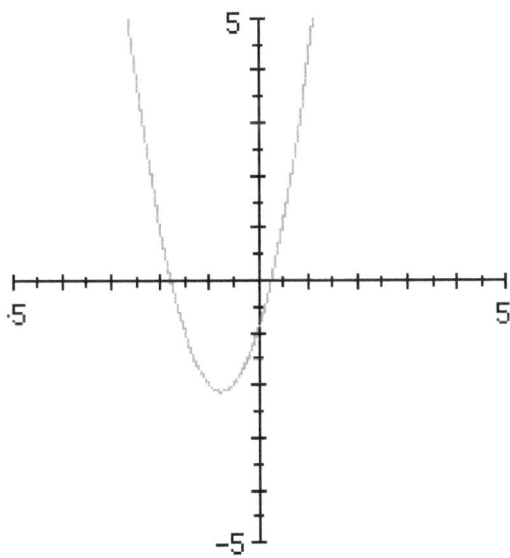

i) $f(x) =$ _____

ii) $f(x) =$ _____

1.4 Interpretation of Graphs

Have you ever tried to describe the physical appearance of another person to someone? If you have a picture of that person, you don't even bother with the verbal description. You just show them the picture. Much the same thing can be said for attempting to describe the behavior of a function between two sets. Since the graph of a function visually displays its key characteristics, the phrase, "A picture is worth a thousand words", has never been more appropriate. The rises, falls, peaks, valleys, and corners of the curve provide the observer important information at a glance which may not otherwise be so apparent by just studying the functional pairing of numbers in a table.

Consider the situation of Louise, owner and manager of a local pet store. When she first opened her store, Louise experimented with the number of hours per week that she kept the store open. Initially it seemed to her that the more hours her store was open, the more profit she would realize. But later it became apparent that if the store was open at times with very few customers, the operation costs – for heating, cooling, lights, employee salaries, etc. – were greater than her revenue and so her profits actually decreased. In order to determine the optimal number of hours to remain open, she plotted her profits as a function of open hours per week for six months while varying the number of hours each week. The graph of her function is given in Figure 1.4.1.

Figure 1.4.1 Profit per week as a function of open business hours.

This picture makes the situation very clear. Louise's weekly profit continues to increase up to a maximum of about $1900 when she remains open for $t = 58$ hours. It then decreases for $t > 58$. Thus, Louise sees that she need not remain open more than 58 hours per week in order to attain the greatest profit. Other information presents itself as well. Suppose, for instance, she is content to earn $1500 per week. By drawing a horizontal line at 1500 on the y-axis, we see that the points of intersection with the graph are (40, 1500) and (70, 1500) meaning that $f(40) = 1500$ and $f(70) = 1500$ if f is the profit function. Thus the curve tells us that she could achieve this value by working either 40 or 70 hours per week. Which do you think she would choose?!

We now wish to get more specific about certain important functional properties that characterize their behavior. Consider next the graphs of $f(x) = x^2$ and $g(x) = x^3$ in Figure 1.4.2.

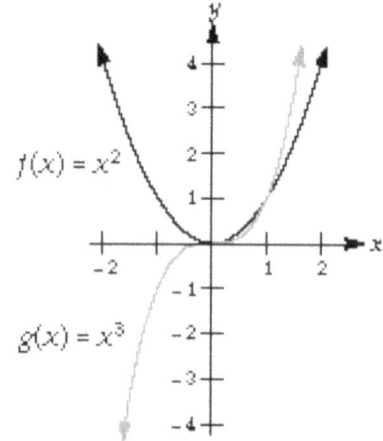

Figure 1.4.2 Graphs of $f(x) = x^2$ and $g(x) = x^3$

What is fundamentally different between these two graphs? Suppose you are a small insect traveling along each curve from left to right. What do you experience? If you are on the graph of $f(x) = x^2$, note that you first travel "down" until you reach $(0, 0)$, at which point you begin to travel "up". This would be similar to a dip in a roller coaster. On the graph of $g(x) = x^3$ however, you first travel up, seem to level off at $(0, 0)$, then continue traveling up again. Hikers often experience this when they realize their path continues upward after reaching a pass. These are the graphical expressions of the terms *increasing* and *decreasing* which we have used previously to describe this behavior except that now we need to associate each region of increase or decrease with the appropriate interval. We would say that the function f is decreasing over $(-\infty, 0]$ and increasing over $[0, \infty)$. Since g is "going up" everywhere we say that it is increasing over $(-\infty, \infty)$ i.e. for all real numbers. Although a calculus textbook may have a more specialized definition, we will say that a function f is **increasing** over an interval if, for every pair of elements $a < b$ in the interval, it is true that $f(a) < f(b)$. Similarly, f is said to be **decreasing** over an interval if, for every $a < b$ in the interval, it is true that $f(a) > f(b)$. The essence of this definition is captured in Figure 1.4.3.

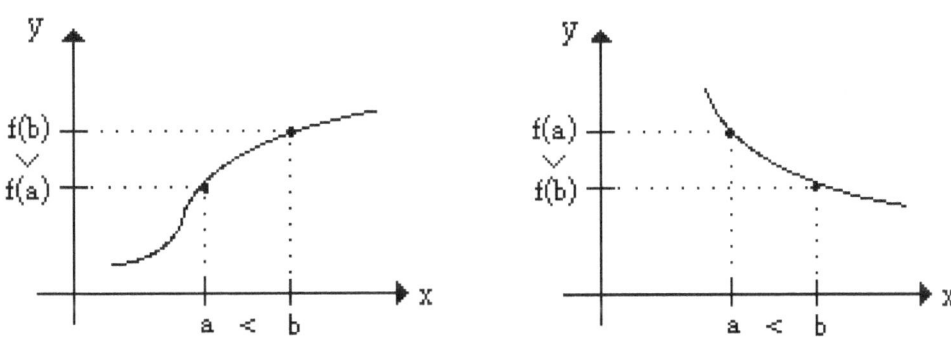

Figure 1.4.3 An increasing function and a decreasing function.

Example 1

The average temperature at any particular geographic location is determined primarily by the latitude and altitude of the location as well as its proximity to large bodies of water. The graph in Figure 1.4.4 depicts mean temperature at sea level as a decreasing function of latitude.

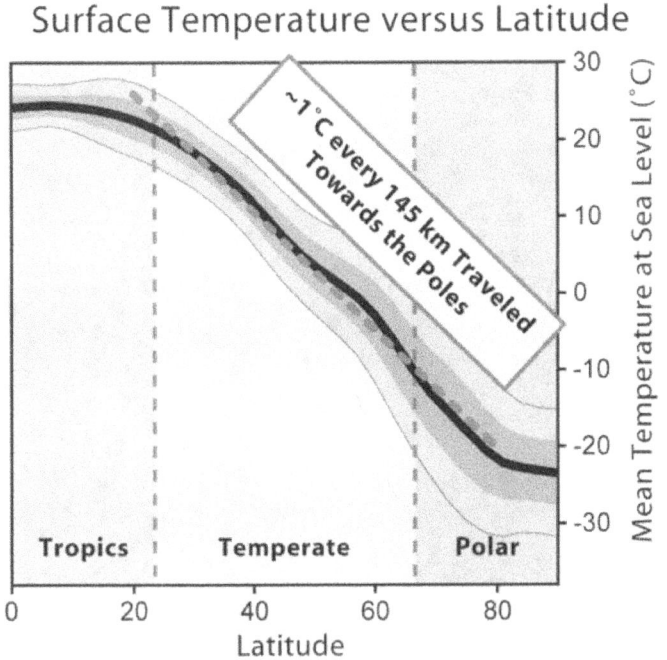

Approximately what is the temperature for most locations between 0 and 20 degrees latitude? By about how much does the temperature decrease as you move from a latitude of 20° to one of 80°? If you wish to live in a region where the mean temperature is above freezing (0° C), below what latitude must you live?

Solution

We can see from the graph that the mean temperature in the tropics remains roughly around 20° C. The temperature decreases from about 20° to about –22°

Figure 1.4.4 Average temperature as a function of latitude.

over the interval 20° to 80° latitude for a net change of about 42°. And by drawing a horizontal line through 0° C, we see that locations between the equator and about 55° latitude have a mean temperature above freezing. ♦

Example 2

The exponential function $f(x) = kb^x$ with base b provides for contrasting examples that depend on the value of the base. If $b < 1$, then we recall from Section 1.2 that the function must get smaller as x advances. For instance, in the case of $g(x) = 3\cdot(0.7)^x$ we see that $g(0) = 3$, $g(1) = 2.1$, $g(2) = 1.47$, and so on. In other words, g is decreasing over $(-\infty, \infty)$. This is reflected in Figure 1.4.4 below by the fact that the graph of g is falling as we scan the picture from left to right. On the other hand, if $b > 1$ then the function will continue to get larger as x gets bigger. For $f(x) = 2\cdot(1.5)^x$ we note that $f(0) = 2$, $f(1) = 3$, $f(2) = 4.5$ and correspondingly, in Figure 1.4.5, the graph of f rises from left to right. It is an increasing function over $(-\infty, \infty)$.♦

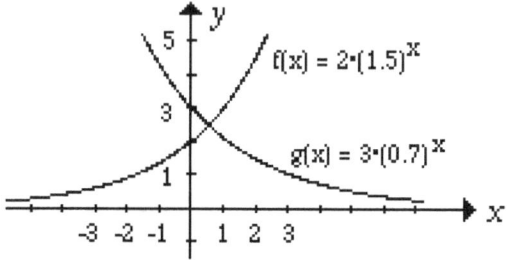

Figure 1.4.5 The graphs of $f(x) = 2\cdot(1.5)^x$ and $g(x) = 3\cdot(0.7)^x$.

We now turn our attention to points where a function *reverses* its direction. We saw in Figure 1.4.2 that the point $(0, 0)$ is a location where the curve of $f(x) = x^2$ switches from decreasing on $(-\infty, 0]$ to increasing on $[0, \infty)$. Since this is the lowest point for all the ordered pairs on the graph, f(0) is the smallest functional value in the range of f and we say that f achieves a **minimum** at $x = 0$ whose value there is $f(0) = 0^2 = 0$. Since the function increases without bound forever, it has no maximum value.

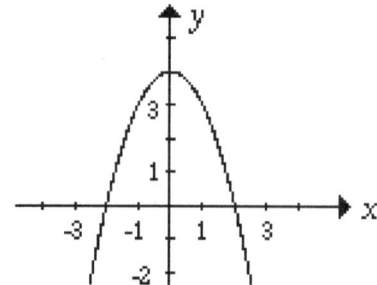

Figure 1.4.6 Graph of $h(x) = -x^2 + 4$.

Alternatively, we see that the opposite is true for the graph of $h(x) = -x^2 + 4$ in Figure 1.4.6. Since $h(x)$ is increasing on $(-\infty, 0]$ and decreasing on $[0, \infty)$, the value $h(0)$ is greater than all the other values in the range of h and we say that h achieves a **maximum** at $x = 0$ whose value is $h(0) = 4 - 0^2 = 4$.

We now return to the general quadratic $f(x) = ax^2 + bx + c$ that we considered briefly in the last section. (The above functions $f(x) = x^2$ and $h(x) = -x^2 + 4$ are both quadratic in form with $b = 0$.) The graph of every quadratic function is a bullet-shaped curve known as a **parabola** and every parabola has a turning point called the **vertex**. The vertex of $f(x) = x^2$ was $(0, 0)$ and that of $h(x) = -x^2 + 4$ was $(0, 4)$. In general, the x-coordinate of the vertex of the graph of $f(x) = ax^2 + bx + c$ is given by $-\dfrac{b}{2a}$ and the corresponding y-coordinate is obtained by substituting this number into the function. The resulting value $f(-\quad)$ must be its minimum or maximum value depending on whether the parabola opens up or down. If $a > 0$, the parabola opens upward and the value is a minimum. If $a < 0$, it opens downward and so a maximum must result. For instance, the vertex of the graph of $f(x) = x^2 - 2x - 3$ must have coordinates:

$$x = -\frac{b}{2a} \quad = -\frac{-2}{2(1)} \quad = 1 \qquad \text{and} \qquad y = f(1) = 1^2 - 2(1) - 3 = -4.$$

Therefore it is located at the point $(1, -4)$. Plotting a few more points quickly leads us to the picture in Figure 1.4.7 where the vertex appears as the bottom of the "valley" of the parabola.

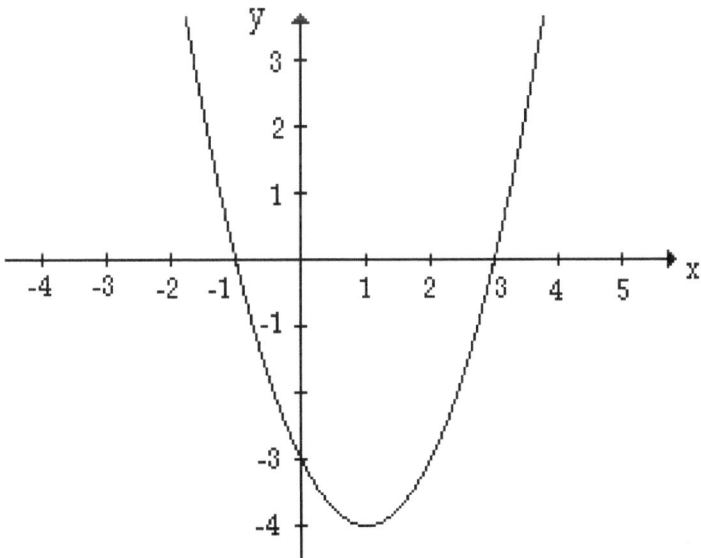

Figure 1.4.7 Graph of $f(x) = x^2 - 2x - 3$

We readily see that $f(x)$ is equal to 0 when $x = -1$ and $x = 3$ by factoring the expression $x^2 - 2x - 3 = (x + 1)(x - 3)$, equating to 0, and solving for x. Values in the domain of a function for which the function equals 0 are known as the **zeroes** of the function. So in this case the zeroes of f are $x = -1$ and 3. We also observe f is decreasing over the interval $(-\infty, 1]$, has a minimum value of -4 for $x = 1$, and is increasing over $[1, \infty)$. The domain for this function is the set of all

61

real numbers $D = (-\infty, \infty)$ while the range includes all numbers greater than or equal to -4 and so consists of the set $R = [-4, \infty)$.

Example 3

A rock is thrown from the edge of a cliff h_0 ft high above ground level with a velocity having a vertical component given initially by v ft/sec. (The vertical component of velocity is the rate at which the rock is gaining or losing height.) The height $h(t)$ of the rock above the ground at the base of the cliff must be a function of the time t after the rock is thrown. (Note that $t = 0$ marks the instant the rock leaves the thrower's hand.) It is a known fact that if we ignore the effects of air resistance this relationship is given by

$$h(t) = -16t^2 + vt + h_0$$

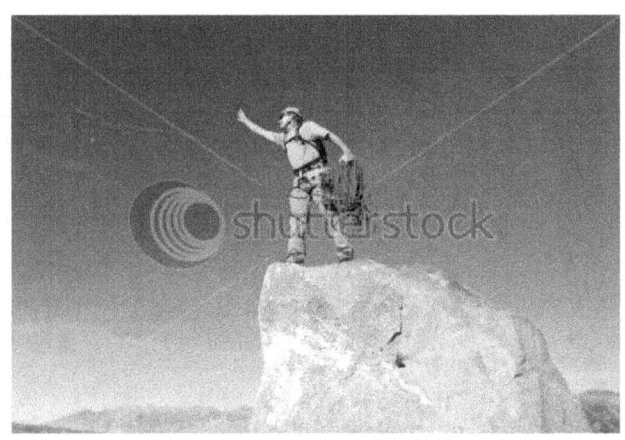

where h is measured in feet and t in seconds. Note that in the absence of gravity the height of the object would be $vt + h_0$. We have seen the $16t^2$ term before in Exercise Set 1.3 as the term which Galileo discovered to be the distance an object drops in a free fall. The negative sign is needed because the positive direction for height is up in this case and gravity is working opposite to the direction of motion of the object. If $v = 48$ ft/sec and $h_0 =$ 64 ft, what is the maximum height attained by the rock? How much time does it take for the rock to hit the ground?

www.shutterstock.com · 19920943

Solution

Since the coefficient -16 of the t^2 term is negative, the graph of this function must be a parabola opening downward. The first coordinate of the vertex is $t = -\frac{b}{2a} = -\frac{48}{-32} = 1.5$ and the second coordinate must be $h(1.5) = -16(1.5)^2 + 48(1.5) + 64 = -36 + 72 + 64 = 100$. The maximum height attained by the rock must be 100 ft. In order to answer the second question, we see that a height of 0 corresponds to the rock hitting the ground and so we factor the quadratic and equate to 0.

$$-16(t^2 - 3t - 4) = 0$$
$$-16(t + 1)(t - 4) = 0$$
$$(t + 1)(t - 4) = 0$$

So the zeroes of h are $t = -1$ and $t = 4$. However, the value of -1 has no meaning here since it would represent a "negative time"! Thus, the domain for h is $D = [0, 4]$ and the range is $[0, 100]$. The rock hits the ground 4 seconds after it was thrown. Noting that this parabola must cross the vertical axis at $h(0) = 64$ gives us enough information to complete a sketch of the graph given in Figure 1.4.8. ♦

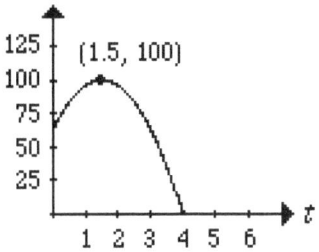

Figure 1.4.8 Graph of $h(t) = -16t^2 + 48t + 64$

Certainly, anybody needing to use a function to draw conclusions about how changes in the independent variable affect the values of the dependent variable would want to have a copy of its graph close at hand. A graph succinctly displays what is important about the relationship between the two variables, no matter what types of physical quantities are involved.

Example 4

Find the maximum and minimum values of $f(x) = 0.75x + 3.5$ for the domain $D = [1, 10]$. Is this an increasing or decreasing function?

Solution

The graph is given in Figure 1.4.9.

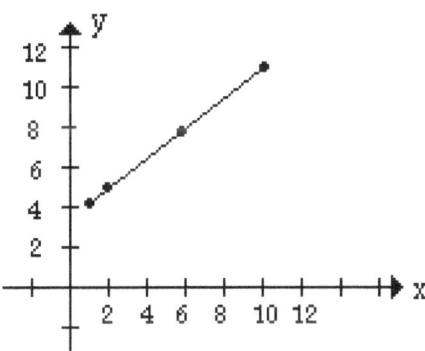

Figure 1.4.9 Graph of $f(x) = 0.75x + 3.5$

We can see that the minimum is given by the y-coordinate of the left endpoint on the line segment, which would be $f(1) = 4.25$. The maximum value of the function over $[1, 10]$ is the y-coordinate of the right endpoint or $f(10) = 11$. Clearly, f is increasing and is never 0 for any x-value in the given interval. Note additionally how the range must include every real number between 4.25 and 11 and so $R = [4.25, 11]$. ♦

Sometimes there is more than one "hill" in the graph of a function, creating a situation where the function may have one or more values which is the largest in its "neighborhood" but not necessarily the biggest value in the entire range of f. When this is the case, we distinguish the functional value of the point at the top of the highest hill by calling it the ***absolute maximum*** and each of those at the top of a minor hill a ***strictly local maximum***. Likewise, the ***absolute minimum*** is the smallest value the function achieves and is positioned at the bottom of the lowest valley while a ***strictly local minimum*** is the bottom of any valley not quite as deep. All of these concepts are on display in the graph of the function f in Figure 1.4.10. Without even knowing the formula for f, we may still glean an abundance of information about its behavior from this picture.

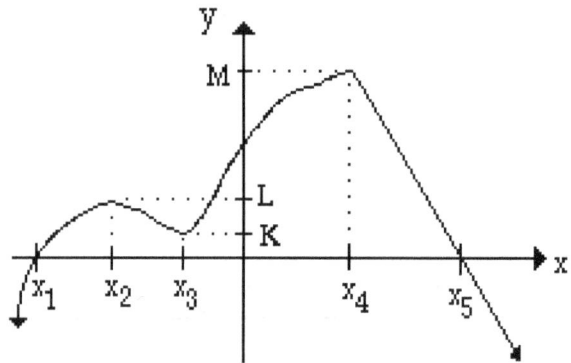

Figure 1.4.10

1. f is increasing over $(-\infty, x_2]$ and over $[x_3, x_4]$. Notice that the curve is "rising" from left to right in each of these intervals.

2. f is decreasing over $[x_2, x_3]$ and over $[x_4 \, \infty)$. Notice that the curve is "falling" from left to right in each of these intervals.

3. f achieves a strictly local maximum at $x = x_2$. The local maximum value is $f(x_2) = L$, since L is the y-coordinate of the point at the top of the hill.

4. f achieves a strictly local minimum at $x = x_3$. The local minimum value is $f(x_3) = K$, since K is the y-coordinate of the point at the bottom of the valley.

5. f achieves an absolute maximum at $x = x_4$. The absolute maximum value is $f(x_4) = M$, since the top of the highest hill has M as its y-coordinate.

6. f never achieves an absolute minimum. The curve plunges down forever as x becomes smaller (more and more negative) and as x becomes larger. If f did have an absolute minimum, it would appear on the graph as the y-coordinate of the deepest valley.

7. The *zeroes* of f are $x = x_1$ and x_5 .

The horizontal axis of many graphs used in the literature of science and business is meant to only contain the domain of the graphed function *as a subset*. It often happens there is just a finite number of values for which the function can be evaluated and once the pairings have been plotted, a smooth curve is extended from point to point to "fill in" the graph. The purpose is only a cosmetic one with no intention to imply that functional values exist at those "filled in" areas. These notions have to do with the mathematically technical concept of *continuity* with which we will not concern ourselves. However, to distinguish such graphs from one for which the horizontal axis does constitute the whole domain, we shall refer to such graphs as **extended graphs**. Extended graphs are widely used to convey information.

Example 5

The distribution of precipitation across the surface of the earth depends, in large part, on the general circulation of the atmosphere. The circulation at any particular location in turn depends on the latitude of the location and so we can plot annual precipitation as a function of latitude for a few select latitudes and then create an extended graph as seen in Figure 1.4.11. At

64

what latitudes does the annual precipitation acquire strictly local maximum and minimum values? At what latitude does the precipitation achieve an absolute maximum?

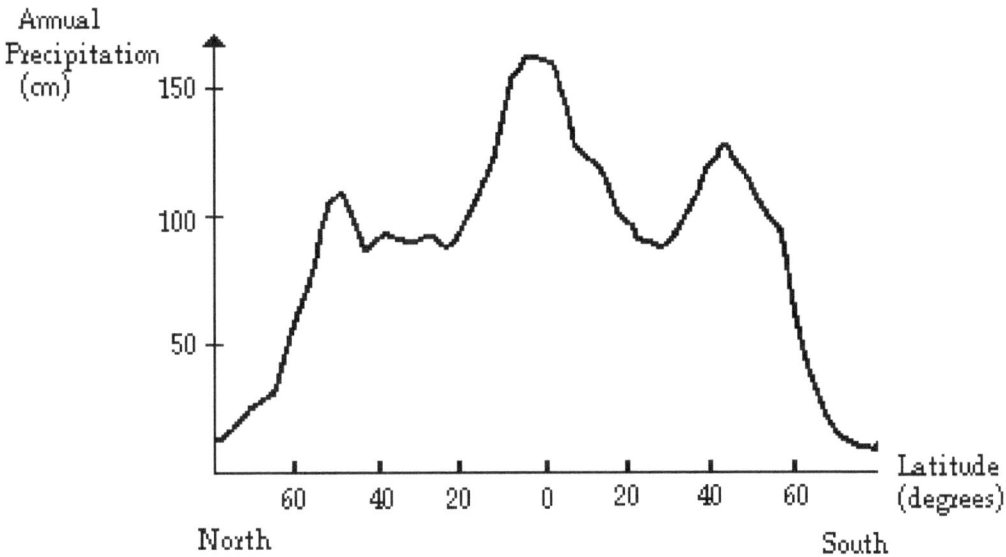

Figure 1.4.11 Annual precipitation as a function of latitude.

Solution

We see from the graph above that there are several strictly local maxima. A prominent one of about 110 cm of precipitation occurs around 50°N latitude and another one of about 120 cm at around 42°S latitude. A strictly local minimum value of about 80 cm occurs for three latitudes between 45°N and 20°N and again at 30°S latitude. An absolute maximum precipitation of over 150 cm occurs just a few degrees north of the equator. ◆

Example 6

Elof is a woodcarver who specializes in making whistles. Elof knows that, up to a point, the more whistles he sells each month the cheaper price he can charge. This, in turn, is good for business. In fact, the price p is a decreasing function of the number x of whistles he sells per month according to $p = -0.08x + 12$. What is the maximum revenue Elof can make per month from the sale of his whistles and how many must he make to achieve this maximum?

Solution

The revenue R realized by Elof is given by the product of the number x of whistles that he sells each month and the price p per whistle. We make R a function of x.

$$\begin{aligned} R(x) &= xp \\ &= x(-0.08x + 12) \\ &= -0.08x^2 + 12x \end{aligned}$$

This is a quadratic function. Since x can only have integer values, the parabola we draw will be an extended graph. The vertex occurs at

$$x = \frac{-b}{2a} \quad = \frac{-12}{-0.16} \quad = 75 \quad \text{and} \quad y = R(75) = -0.08(75)^2 + 12(75) = 450.$$

We equate the factored version of $R(x)$ to 0 to find the zeroes of the function.

$$x(-0.08x + 12) = 0 \qquad \text{for } x = 0 \text{ and } x = \frac{12}{0.08} \quad = 150$$

The extended graph appears in Figure 1.4.12.

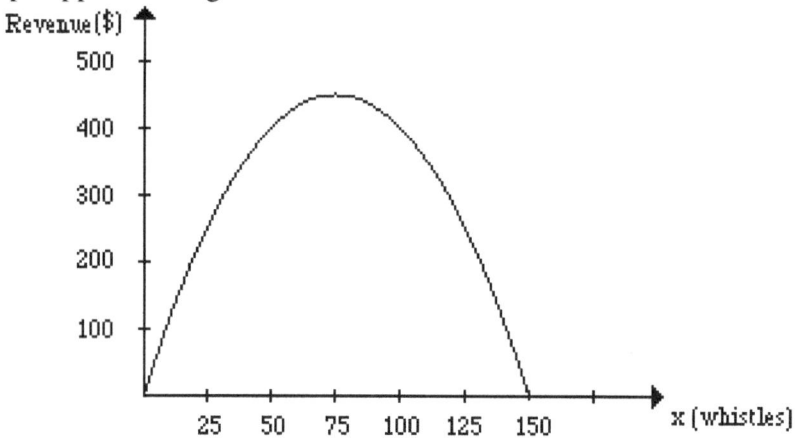

Figure 1.4.12 Elof's revenue as a function of the number of whistles.

If Elof carves and sells 75 whistles each month, he will make a maximum revenue of \$450. If he makes more than 75 whistles per month, his revenue will decrease. In fact, if he makes 150 whistles, his revenue will drop to nothing! ♦

The above example illustrates one the basic concepts of conventional economic theory known as the ***Law of Demand.*** The price of a product decreases as the quantity which consumers demand (or may purchase at that price) increases. Note this matches our intuitive observation that as price increases, the demand decreases. So this law may be stated:

Demanded quantity is a decreasing function of the unit price.

A related relationship is the ***Law of Supply.*** This embodies the principle that the manufacturers of a particular item will tend to devote more resources to its production as the price increases. This is concisely stated as:

Supplied quantity is an increasing function of the unit price.

By graphing both the demand curve and the supply curve for the same product on the same set of axes, we can located the point of intersection (p_0, q_0) of the two graphs known as the ***equilibrium point***. Since this identifies the price at which the demand is equal to the supply, the price p_0 is called the ***equilibrium price*** and the quantity q_0 is called the ***equilibrium quantity***. These concepts are illustrated in Figure 1.4.13. Note that for any unit price greater than p_0, the supply exceeds the demand resulting in a *surplus* of the product. On the other hand, if the unit

66

price is less than p_0, the supply is less than the demand creating a situation called a product *shortage*.

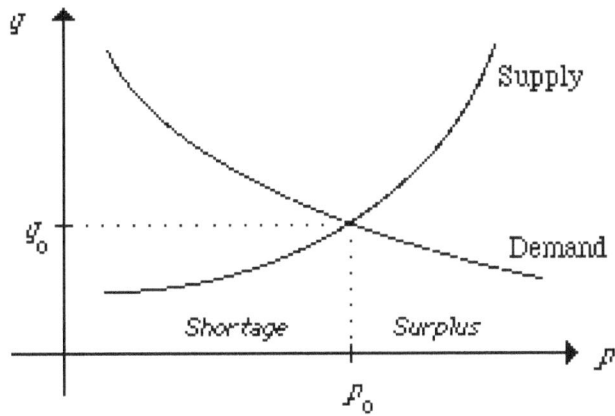

Figure 1.4.13 Supply and demand curves graphing quantity q as functions of price p.

We see that graphs are capable of conveying large amounts of information quickly and succinctly. They are indispensable tools in modern society for efficiently charting relationships between varying quantities. Mastery of the ability to make an accurate analysis of a variety of different types of graphs improves your ability to make decisions concerning your job, your home, and your life.

Exercise Set 1.4

Use each of the following graphs to determine:
   i) the domain and range
   ii) over what intervals the function increases or decreases
   iii) maximum or minimum values and where they occur
   iv) the zeroes of the function

1.

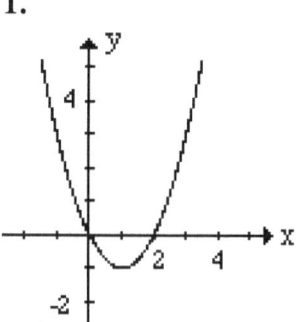

2.

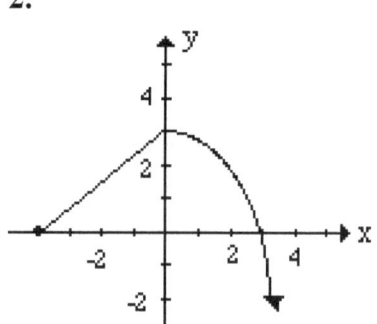

3.

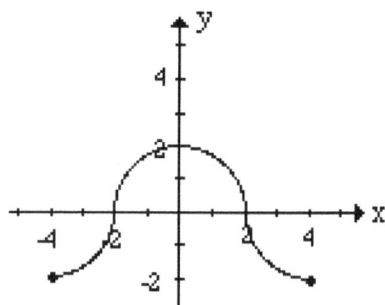

4.

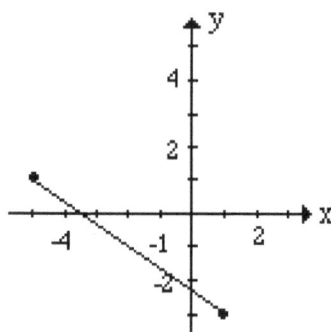

5.

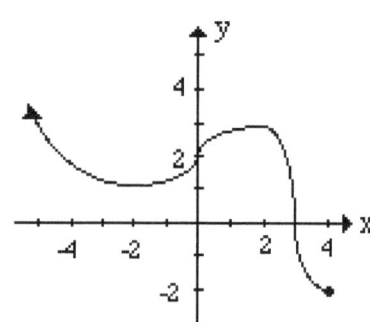

6.

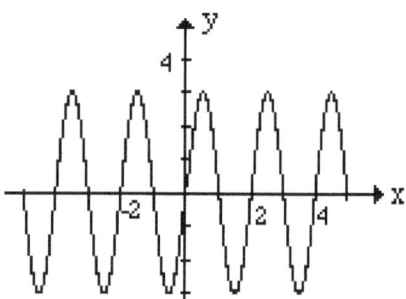

7.

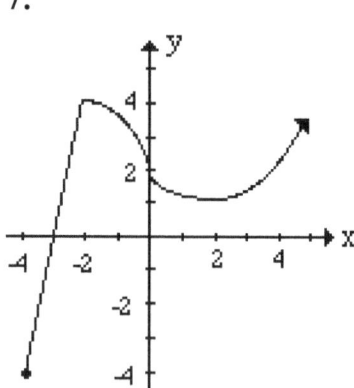

8.

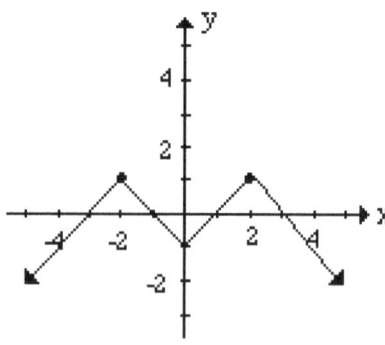

9.

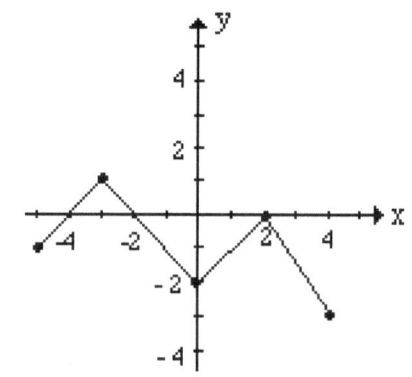

10. Figure 1.4.12 gives the annual precipitation as a function of latitude. Is the annual precipitation increasing or decreasing between 0° and 30° S? Above what northern latitude is the precipitation less than 50 cm?

Graph the following functions. In each case, use the graph to locate any maxima, minima, or zeroes and determine where the function increases or decreases.

11. $f(x) = 3x + 12$ over $[1.1, 8.4]$

12. $f(x) = 8x - 5$ over $[-3, 6.5]$

13. $g(t) = -1.6t + 7$ over $[2, 3]$

14. $h(x) = -2.5x + 4$ over $[1.3, 5]$

15. $f(x) = -x^2 + 4$

16. $g(x) = x^2 + 1$

17. $f(x) = -x^2 - 2x$

18. $f(t) = t^2 + 4t$

19. $h(x) = x^2 - 5x + 7$

20. $h(x) = 3x^2 - 4x + 2$

21. $g(t) = -t^2 - 2t + 8$

22. $f(x) = 2x^2 - 9x - 5$

23. $h(t) = 3t^2 - 5t - 2$

In the absence of air resistance, the height of an object (in feet) thrown from a cliff h_o feet high having an initial vertical component of velocity v ft/sec is a function of time t (in sec) given by

$$h(t) = -16t^2 + vt + h_o.$$

Find the maximum height attained by the object and the elapsed time it takes for the object to strike the ground given each of the following values for v and h_o .

24. $v = 16$ ft/sec ; $h_o = 96$ ft

25. $v = 32$ ft/sec ; $h_o = 128$ ft

26. $v = 48$ ft/sec ; $h_o = 160$ ft

27. $v = 96$ ft/sec ; $h_o = 112$ ft

Use a graphing calculator to estimate any maxima, minima, or zeroes to the nearest tenth.

28. $f(x) = 2x^3 - x^2 - 2x + 3$

29. $f(x) = x^4 - 2x^2 + 3$

30. $g(x) = x^4 + 3x^2 - 4x + 1$

31. $g(x) = 2x^3 - 5x^2 + 2$

32. $h(x) = -x^3 + 2x - 1$

33. $h(x) = -2x^5 + 5x^3 - 3x + 1$

34. It is more realistic to consider air resistance when obtaining the equation of the height of a thrown object as a function of time. If we assume the decelerating effect of air resistance to be proportional to the velocity of the object with a proportionality constant of 0.4, then the height of the object thrown up from the ground ($h_o = 0$) with an initial speed of 25 ft/sec is given by

$$h(t) = -80t - 262.5\, e^{-0.4t} + 262.5$$

The graphs of the two height functions with and without air resistance are given below. Estimate the responses to the following questions using these graphs.

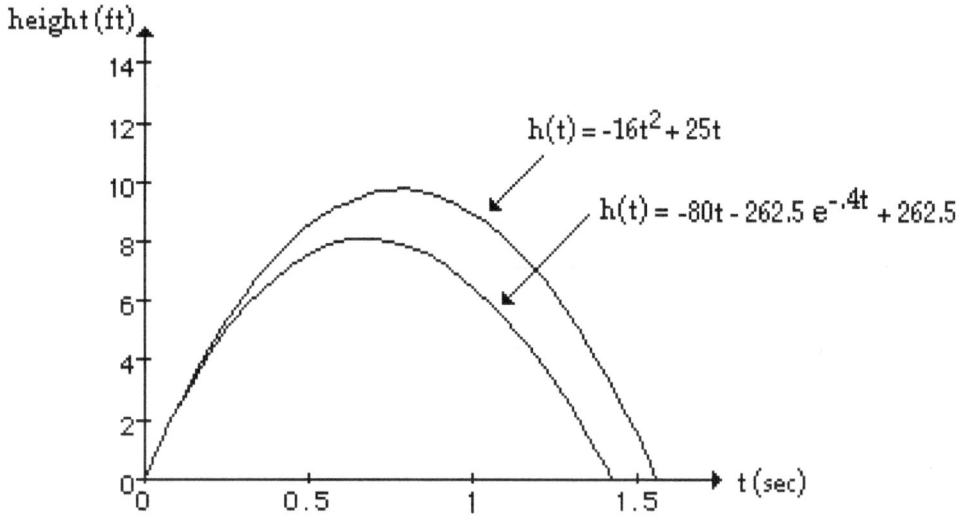

i) What are the maximum heights attained with and without the air resistance assumption?   Which assumption yields the bigger maximum?

ii) At what times are these maximum heights achieved?

iii) Under which assumption is the object in the air for the longest amount of time? What amount is that?

iv) For what interval of time is $h(t) = -16t^2 + 25t$ still increasing while $h(t) = -80t - 262.5\, e^{-0.4t} + 262.5$ is decreasing?

35. Assume the pressure under water exerted by the ocean changes linearly with the depth. If the pressure at sea level is 1 atm (atmospheres) and the pressure at 100 ft under water is 4 atm, find the water pressure as a function of depth. Use it to compute the maximum pressure a submarine designed to dive down to 1500 ft would have to endure.

36. Fahrad has learned that the monthly revenue R he makes from his doughnut shop is a function of the number x of doughnuts he bakes each day according to
$$R(x) = -0.02x^2 + 9x$$
where R is measured in dollars. Graph the function.

37. Use the graph of the function in the previous problem to find the maximum revenue and the number of doughnuts Fahrad must bake each day in order to achieve that maximum.

38. The *photosphere* is the name given to that layer of the sun that is visible to the human eye. At the photosphere, the temperature is about 6000°K. Up to about 400 km above the photosphere, the temperature (in degrees Kelvin) decreases according to $T(x) = 6000 - 5x$ where x = height above the photosphere in kilometers. What is the minimum value of the temperature over the domain $[0, 400]$?

39. A land developer wishes to form a rectangular lot having the greatest possible area and such that the sum of the boundaries is equal to 600 ft. What are the length, width, and area of this rectangle? (*Hint:* If x is the length, then $300 - x$ must be the width.)

40. *Whoopee Toys* is marketing an action figure called the Rodent Warlock. The unit price has been determined to be $p = 6 - 0.25x$ where x is the number of warlocks to be sold. Find the maximum possible revenue and the number of warlocks which Whoopee Toys should produce in order to achieve that maximum. (*Hint*: See Example 6.)

41. The figure below gives the supply and demand curves for (hypothetical) bushels of soyoats. Each curve is a graph of quantity (bushels per week) as a function of price (dollars).
  (i) Which function is increasing? decreasing?
  (ii) What are the equilibrium price and quantity?
  (iii) Over what interval of prices will there be a surplus of soyoats? Over what interval of prices will there be a shortage of soyoats?

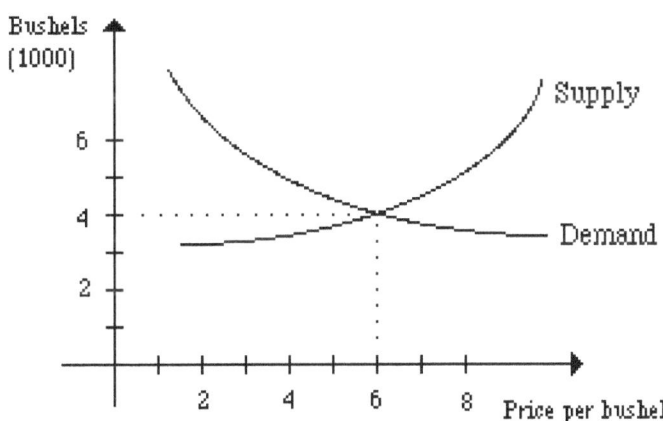

42. Steven G. Monk is a mathematician at the University of Washington who researches how students learn mathematical concepts. In an experiment, he presented the following graph to a student giving the *velocities* of two cars (Red and Blue) as functions of time. He asked the question, "Between 5 and 15 seconds, is car Blue catching up to car Red or are they getting farther apart?" How would you answer that question? Justify your answer. [Results presented in an address given at the annual Joint Mathematics Meetings in Cincinnati, Ohio, January, 1994.]

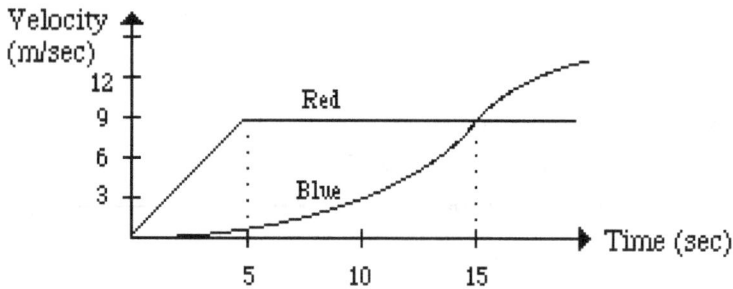

43. The intensity of radiation from a heated object depends on the temperature of the object and the wavelength of light being emitted. The graphs below show the intensity as a function of wavelength (μm) for several temperatures in degrees Kelvin. (1 μm = 10^{-6} meter)

i) For each temperature, estimate the wavelength for which the object radiates at maximum intensity, usually denoted by λ_{max} .

ii) Is λ_{max} an increasing or decreasing function of temperature? (This fact is known as Wein's Law.)

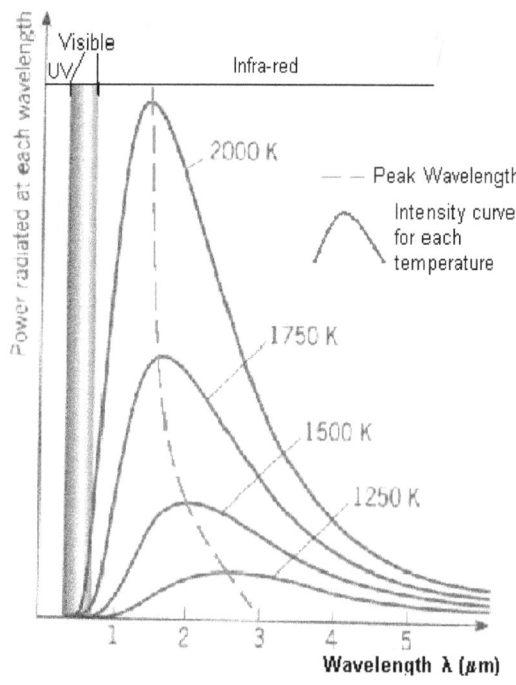

72

44. The following graph represents the profitability of a product at different stages of its life cycle. Placing a product on this timeline suggest certain strategies for keeping sales high. For example, if sales sag even though the product is thought to be in a growth period, then action may be taken such as wider distribution or discounts for a large volume of sales. What do you think might the units of time for a product such as a car? A type of sneaker? Seasonal clothes?

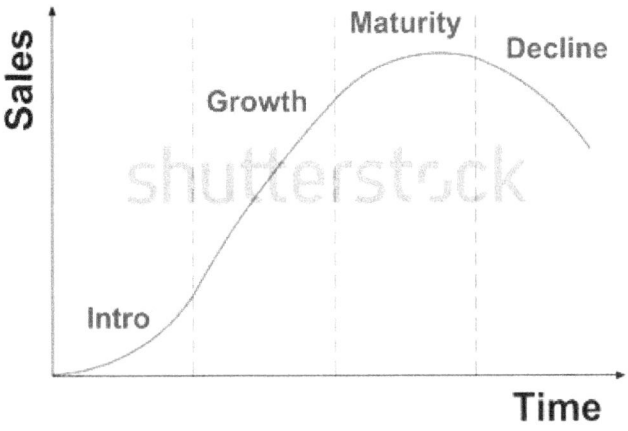

Product Life Cycle

45. An *oscillator* is any system that exhibits periodic behavior. Oscillators are sudied by biologists since they are found extensively in living organisms: pacemaker cells in the heart, insulin secreting cells in the pancreas, and neural networks in the brain. The pendulum mentioned in Section 1.1 is another example of an oscillator. The two graphs labeled 1 and 2 on the axes below represent the displacement from rest and the velocity of the associated moving pendulum as functions of time. Which graph corresponds to displacement? to velocity? (Consider the vertical axis to represent either distance or velocity units.)

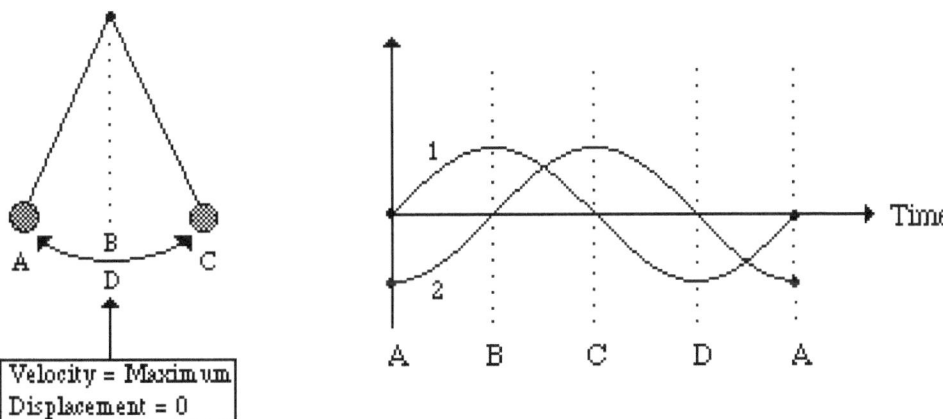

46. According to the graph of given at the right, about what year will the world population reach 8 billion? In what year will it reach 10 billion? In 1950, the developing countries constituted about what percentage of the total? About what percentage is predicted for 2050?

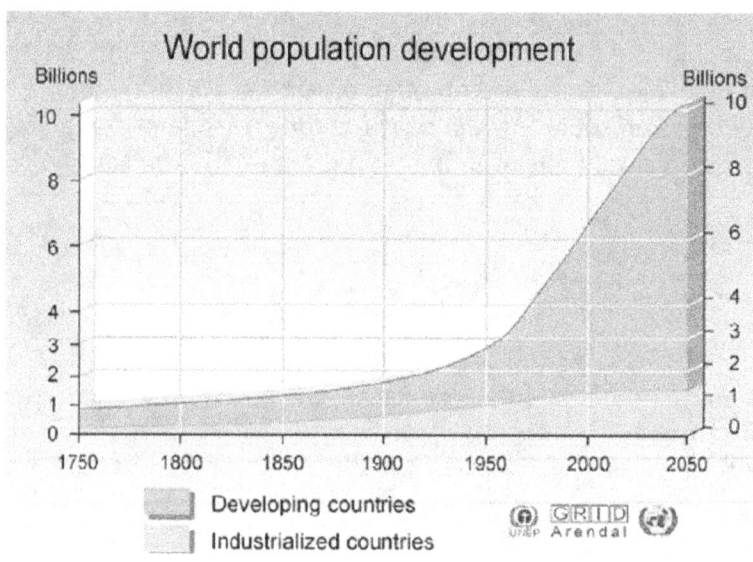

Chapter Glossary

absolute maximum The largest value a function attains in its range.

absolute minimum The smallest value a function attains in its range.

action The steps taken by a function to process an element x in its domain.

algorithm Sequence of specific steps to compute a value or solve a problem.

base The factor by which an exponential function is consistently multiplied per unit change in the independent variable. It is denoted by b in the general formula for the function. It is necessary for $b > 0$ and $b \neq 1$.

Cartesian coordinate system A system of perpendicular number lines called axes used to graph functions. Also referred to as rectangular coordinate system.

cryptology The study of the techniques used for creating secret codes.

decreasing A function f is said to be decreasing over an interval if $f(a) > f(b)$ whenever $a < b$ for any pair of numbers a, b in the interval.

dependent variable The variable in a two-variable relationship whose value depends on the value assigned to the independent variable.

domain The set of elements which are the inputs to a particular function.

equilibrium price The price for which the supply quantity is equal to the demand quantity.

equilibrium quantity The quantity attained by both the supply and demand functions at the equilibrium price.

extended graph The graph obtained by extending a smooth curve between points on the graph of a function whose domain is only a subset of the interval displayed on the x-axis.

exponential function A function of the form $f(x) = kb^x$.

formula The precise expression for a function which describes its action in terms of the independent variable .

function A relationship between two sets which assigns to every element in the first set (domain) exactly one element in the second set (range).

graph of a function f The set of all ordered pairs (x, y) plotted on a Cartesian coordinate system in which the y-coordinate is equal to $f(x)$ and x is a member of the domain of f.

image The specific functional value in the range of a function associated with a particular value in the domain of the function.

increasing A function f is said to be increasing over an interval if $f(a) < f(b)$ whenever $a < b$ for any pair of numbers a, b in the interval.

independent variable The variable in a two-variable relationship or a function which is thought of as varying freely among the allowed values.

linear function A function of the form $f(x) = mx + k$.

natural exponential function An exponential function with base given by a power of e.

nonlinear function Any function which is not linear.

parabola The graph of a quadratic function.

quadratic function Any function of the form $f(x) = ax^2 + bx + c$.

range The set of elements which are the outputs of a particular function.

rate The change in the functional value of a linear function per unit change in the independent variable.

strictly local maximum The largest value of a function for an interval of "nearby" domain values. It is not an absolute maximum. It appears on the graph as the y-coordinate of a point at the top of a minor hill.

slope Characteristic of a linear graph computed by   for any two points (x_0, y_0) and (x_1, y_1) on the line. It is numerically equal to the rate.

strictly local minimum The smallest value of a function for an interval of "nearby" domain values. It is not an absolute minimum. It appears on the graph as the y-coordinate of a point at the bottom of a minor valley.

vertex The point on a parabola at which the curve turns around. The x-coordinate of the vertex is given by $x = -b/2a$ where $f(x) = ax^2 + bx + c$ is the associated quadratic function.

vertical line test A graph represents a function of the variable associated with the horizontal axis if no vertical line intersects the graph more than once.

zero of a function A value r of a function f such $f(r) = 0$.

Chapter 1 Review Test

1. Suppose we wish to assign to each consonant in the alphabet the closest vowel. For example, h → i and q → o. Would this be a function? Why or why not?

2. Assume everyone in a room has one of the distinct eye-colors blue, green, brown, or hazel. If you wish to define a function between the set of these eye-colors and the set of people in this room, which set must be the domain and which set must be the range?

3. The probability of heart disease of a person is an increasing function of the average amount of fat consumed daily by that person. Explain what this means.

4. The French mathematician-soldier credited with introducing the concept of a graph by using a rectangular coordinate system was _____ .

5. The gravitational acceleration a imparted by a body of mass M to any object is inversely proportional to the square of the distance r of the object from the center of the mass. In symbols,
$$a = GM / r^2$$
where G is the gravitational constant. Is a an increasing or decreasing function of r?

6. Given the function $f(x) = 350 + 6.2$ _____, compute $f(28)$, $f(52)$, and $f(70)$ to the nearest tenth. Is f an increasing or decreasing function? (For $x \leq 77$.)

7. A milking machine at a local dairy was originally purchased for 7200 dollars. If it depreciates linearly in value according to the following table, find the expression for the value $V(t)$ as a function of time t and use it to forecast the value of the machine in 10 years.

Time (years)	Value (dollars)
0	7200
1	6900
2	6600
3	6300
4	6000
5	5700

8. Construct a formula for each of the functions which make the assignments indicated in the following tables of values.

(i)

x	f(x)
5	11
10	26
15	41
20	56
25	71

(ii)

x	g(x)
−2	5/16
−1	5/4
0	5
1	20
2	80

9. Which of the following graphs do not represent functions?

(i)   (ii)   (iii)

10. Graph each of the following functions.
      i) $f(x) = 2x + 3$ ii) $g(x) = (0.75)\, 2^x$

11. Give the domain, range, absolute maximum value, and absolute minimum value for each of the following functions.

(i)   (ii)

12. The function $h(t) = -16t^2 + 64t + 80$ gives the height (in feet) of an object thrown from a cliff 80 feet high with an initial speed of 64 ft/sec as function of time t (in sec). Graph the function and find the maximum height attained by the object and the time it takes for the object to hit the ground.

13. Examine the following graph of the function f to answer the following true/false questions.

(i) $f(-3) < f(3)$ (iv) If $x > -1$, then $f(x) > 0$

(ii) $f(0) = 4$ (v) $f(-2)$ is the absolute minimum value of f.

(iii) $f(-7) < 0$

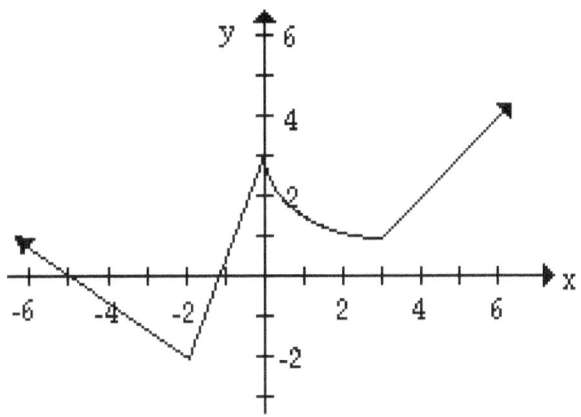

14. Fill in the blanks based on the above graph of the function *f*.

(i) *f* is increasing over the intervals _____ and _____ .

(ii) *f* has a strictly local minimum value of _____ at *x* = _____ .

(iii) *f* has a strictly local maximum value of _____ at *x* = _____ .

(iv) The zeroes of *f* occur at *x* = _____ .

15. Suppose you deposit $3200 in a savings account at a bank offering an annual interest rate of 4.7% compounded annually. Form the expression that gives the amount of money accrued as a function of time t in years. What amount of money will you have in that account in 2 years? 5 years?

16. The population of Gotham City is currently 217,000 and growing at an annual rate of 1.6%. If the growth is modeled by the natural exponential function, what will be the population of the city in 4 years? 8 years?

17. Create a formula for each function described by the following procedures.

(i) 1. Pick any number.
2. Add 3 to the number and then square it.
3. Multiply the result by 6.

(ii) 1. Pick any number.
2. Multiply the number by 7 and then subtract 4.
3. Take the cube root of the result.

Chapter 3
Logic and Computer Science

And this shows that intuition can sometimes get things wrong. And intuition is what people use in life to make decisions. But logic can help you work out the right answer.

Mark Haddon
The Curious Incident of the Dog in the Night-time

In Mark Haddon's wonderful book about the adventure of an autistic teenager, another window on the nature of logical reasoning is thoughtfully explored. Similarly, the classic tales of *Alice in Wonderland* and *Through the Looking Glass* by Lewis Carroll are generally considered to be wonderful works of fantasy written for the amusement of children. However, few people realize that Carroll was the pen name of a professor of mathematics and logic at Oxford, Charles Dodgson (1832–1898). Although Dodgson's primary purpose in both of these books was entertainment, many of the adventures of the main character Alice are enlivened with notions involving *logic*. In addition to his many humorous works, Dodgson wrote several serious books in his field including *Symbolic Logic* in which he wrote,

> It [symbolic logic] will give you clearness of thought – the ability to *see your way* through a puzzle – the habit of arranging your ideas in an orderly and get-at-able form – and, more valuable than all, the power to detect *fallacies*, and to tear to pieces the flimsy illogical arguments which you will so continually encounter in books, in newspapers, in speeches, and even in sermons . . .

Logic – the study of correct reasoning – is the glue which binds together any mathematical framework and, in turn, any science or applications supported by that mathematics. Logic dictates the validity of arguments and therefore must be the essential tool not only of scientific realms, but also of philosophy, law, and any other arena which requires clearness of thought and precise reasoning. We begin by introducing the basic mathematical building blocks which we will later combine with the rules of logic in order to systematically evaluate the truth of an argument. We will then show how logic is used in the construction of a computer program by featuring certain types of algorithms and decision-making. This chapter uses ideas which tie together a number of concepts previously presented in Chapters 1, 3, and 4.

3.1 Statements and Connectives

One of the grandest examples of pure logic is the 2000-page book *Principia Mathematica* written by the British mathematicians **Alfred North Whitehead** (1861–1947) and **Bertrand**

Russell (1872–1970) over a period of ten years and published between 1910 and 1913. This book begins by defining a few terms and listing a small number of axioms or assumed facts and proceeds to *deductively* reach a large number of conclusions concerning the foundations of mathematics. Many consider this towering work to be the greatest contribution to the science of logic since the time of Aristotle. As ***deduction*** is the process of correctly reaching a conclusion from a given set of initial axioms or statements, it is no accident that both Russell and Whitehead were also two of the 20th century's premier philosophers. After losing his son in combat in World War I, Whitehead turned his attention in earnest to philosophical musings with the publications of *An Enquiry Concerning the Principles of Natural Knowledge* (1919) and *The Concept of Nature* (1920). After a successful thirty year stay as a mathematician at Trinity College, Cambridge, Whitehead finished his career as professor of philosophy at Harvard where he wrote *Science and the Modern World* (1925) – a highly acclaimed study of the effects of modern science on Western

culture. Russell, who was a student of Whitehead at Cambridge, was one of history's great pure thinkers. In addition to mathematics, he also wrote extensively on scientific, philosophical, political, and social concerns. Among these are: *The Analysis of Mind* (1921), *The Prospects of Industrial Civilization* (1923), *The ABC of Relativity* (1925), *An Outline of Philosophy* (1927), *The Scientific Outlook* (1931), *Education and the Social Order* (1932), *Power* (1938), *Human Knowledge* (1948), and various other works. He received a multitude of distinctions for his achievements including the Nobel Prize in Literature in 1950.

Figure 3.1.1 Bertrand Russell (1872–1970)

In disciplines such as mathematics and philosophy often you begin with a group of given *objects*, define *operations* on those objects, and then obtain conclusions by rigorously applying those operations to those objects. You have been utilizing this concept since childhood. In the game of "Checkers", for instance, the objects are the red and black pieces and the operations are the rules for the movement of those pieces. Every game concludes after a finite sequence of moves. In logic our objects are ***statements*** for which we will define four operations, known as ***connectives,*** for combining statements. Our conclusions about these combinations will then stem from their ***logical value***. A ***statement*** is a declarative sentence which is has exactly one ***logical value*** of either true or false.

Example 1
Which of the following are statements?
- i) Go to the end of the line.
- ii) Tomorrow is June 21.
- iii) Are you going to the concert tonight?
- iv) That snowmobile can go faster than 45 mph.
- v) Lions are beautiful animals.
- vi) This sentence is false.

Solution

    i) This is a command and is neither true nor false. It is not a statement.

    ii) This is a statement – tomorrow either is or is not June 21.

    iii) This is a question and is not a statement.

    iv) This is a statement – the snowmobile either can or cannot go faster than 45 mph.

    v) It is a matter of opinion whether lions are beautiful. Therefore, this is not a statement.

    vi) This is not a statement because it cannot be given a consistent logical value. If you consider it to be true then it claims itself to be false, but if you consider it to be false then it must be true! (Such an undecidable sentence is sometimes called a **paradox**. See Exercise 15.) ◆

Next we define a *function L* which assigns to any statement p its logical value. Note that this is a well-defined function because a statement must be either true or false but it cannot be both. The domain of L consists of the set S of all statements and the range is the two-element set {True, False}.

p	$L(p)$
$6 + 7 = 13$	True
Mount Everest is the tallest mountain on Earth.	True
All bears are black.	False
$10^3 = 1000$	True
George Bush was a U.S. Senator.	False
$39 + 56 < 41$	False

Sometimes the logical value of a statement can change. The value of "Today is Friday" depends on the current day of the week. If one or more real number variables is used in the declaration of a statement, then we call it an **algebraic** statement. A algebraic statement depends on the current value(s) of the variable(s) involved, and hence so does its logical value.

Example 2

Let p represent: $x + 100 > 200$. Then p changes precisely when x changes.

x	p	$L(p)$
27	$127 > 200$	False
75	$175 > 200$	False
100	$200 > 200$	False
101	$201 > 200$	True
116	$216 > 200$	True ◆

Compound statements are formed by combining a pair of given statements using a key word called a **connective**. (Statements which contain no connectives as in the above examples are known as **prime** statements.) For instance, "Jim is driving the car and Becky is a passenger," is a compound statement formed by linking the two prime statements, "Jim is driving the car," and "Becky is a passenger" with the connective "and". The four main connectives are "and" (**conjunction**), "or" (**disjunction**), "If . . . then" (**conditional**), and "if and only if" (**biconditional**). In addition, one may reverse the logical value (from true to false or from false to true) of any statement by use of the word "not". This is called the *negation* of the original statement. The following are examples of compound statements.

"Jim is driving the car or Becky is a passenger."
"Jim is driving the car and Becky is not a passenger."
"It is not the case that Jim is driving the car or Becky is a passenger."
"If Jim is driving the car, then Becky is a passenger."
"If Jim is not driving the car, then Becky is a passenger."
"If Becky is a passenger, then Jim is driving the car."
"Jim is driving the car if and only if Becky is a passenger."
"Jim is driving the car if and only if Becky is not a passenger."

It is important to note the difference between the conditional and the *bi*conditional. The biconditional stands for two conditionals at the same time. In the statements above, the biconditional, "Jim is driving the car if and only if Becky is a passenger," is completely equivalent to the pair of conditionals, "If Jim is driving the car, then Becky is a passenger," combined with, "If Becky is a passenger, then Jim is driving the car." The notion of equivalent statements will be further discussed in the next section.

In order to streamline the logical analysis of compound statements, letters (such as p, q, and r) are used to represent the prime statements while connectives and negation are denoted by the symbols shown in the table below.

Name	Word	Symbol
Conjunction	and	$\wedge$
Disjunction	or	$\vee$
Conditional	If . . . then	$\Rightarrow$
Biconditional	if and only if	$\Leftrightarrow$
Negation	not	$\sim$

Figure 3.1.2 The symbols used for connectives and negation.

Example 3
   Let p represent, "I mow the lawn on Friday."
   Let q represent, "I play golf on Saturday."

Express the following as English statements:
   i) $p \wedge q$ ii) $p \vee q$ iii) $\sim p$ iv) $p \Rightarrow q$

   v) $p \Leftrightarrow q$ vi) $\sim p \wedge \sim q$ vii) $\sim(p \wedge q)$ viii) $\sim q \Rightarrow \sim p$
Solution
   i) I mow the lawn on Friday and I play golf on Saturday.

ii) I mow the lawn on Friday or I play golf on Saturday.

iii) I do not mow the lawn on Friday.

iv) If I mow the lawn on Friday, then I play golf on Saturday.

v) I mow the lawn on Friday if and only if I play golf on Saturday.

vi) I do not mow the lawn on Friday and I do not play golf on Saturday.

vii) It is not the case that I mow the lawn on Friday and I play golf on Saturday.

viii) If I do not play golf on Saturday, then I do not mow the lawn on Friday. ♦

The English language often allows for several ways to express the same thought. Care must be used when translating a given English compound statement into symbols.

Example 4

Translate each of the following into symbols.

i) If a boy is Swedish, then he is blond.

ii) Jennifer ran fast, yet she lost the race.

iii) All redwood trees grow to be over 100 feet tall.

iv) Neither Mike nor Alfreda play poker.

v) Dark clouds are present whenever you see lightning.

vi) On Sundays, Albert plays either bridge or chess.

vii) Barbara likes psychology, but she does not like statistics.

viii) That shirt is clean if and only if Alex washed it.

Solution

i) Using p for, "A boy is Swedish," and q for, "He is blond," we write: $p \Rightarrow q$.

ii) This statement is equivalent to: "Jennifer ran fast and she lost the race." Using p for, "Jennifer ran fast," and q for, "She lost the race," we write: $p \wedge q$.

iii) This statement can be re-written as: "If a tree is a redwood, then it grows to be over 100 feet tall." Using p for, "A tree is a redwood," and q for, "A tree grows to be over 100 feet tall," we write: $p \Rightarrow q$.

iv) This statement can be re-written as: "Mike does not play poker and Alfreda does not play poker." Using p for, "Mike plays poker," and q for, "Alfreda plays poker," we write:

$\sim p \wedge \sim q$. Observe that this is different from: "Either Mike or Alfreda does not play poker." This would be translated as: $\sim p \vee \sim q$.

v) This statement is equivalent to: "If you see lightning, then dark clouds are present." Using p for, "You see lightning," and q for, "Dark clouds are present," we write: $p \Rightarrow q$.

www.shutterstock.com · 3439881

vi) Using p for, "Albert plays bridge on Sundays," and q for, "Albert plays chess on Sundays," we write: $p \vee q$.

vii) The connective "but" is equivalent for our purposes to "and". Using p for, "Barbara likes psychology," and q for, "Barbara likes statistics," we write: $p \wedge \sim q$.

viii) Using p for, "That shirt is clean," and q for, "Alex washed the shirt," we write: $p \Leftrightarrow q$. ♦

In the conditional statement $p \Rightarrow q$, p is known as the **antecedent** and q as the **consequent**. The conditional has a plurality of forms of English expression. In part (iii) of the example above, the statement had to be re-written using our typical conditional language. This will be true for such categorical statements beginning with the word, "all." In part (v), the antecedent did not come first in the sentence, but instead it followed the word, "whenever." Another variety is sentences like, "The Dow Jones average increases every time interest rates drop." The antecedent here is, "Interest rates drop," while the consequent is, "The Dow Jones average increases."

We can continue this process by combining compound statements with other statements, but we must be careful to include parentheses to indicate which pairs of statements are to be combined first. As in algebra, an expression contained within a pair of parentheses is considered to be a single unit possessing a meaning independent of the rest of the expression.

Example 5
Let p represent, "Philadelphia wins their last game."
Let q represent, "Pittsburgh wins their last game."
Let r represent, "Chicago wins the pennant."

Translate each of the following into symbols. (Teams cannot tie.)
   i) Either Philadelphia wins their last game and Pittsburgh loses their last game or Chicago wins the pennant.
   ii) It is not the case that Chicago wins the pennant if Philadelphia wins their last game.
   iii) If Philadelphia loses their last game and Pittsburgh wins their last game, Chicago wins the pennant.
   iv) Chicago wins the pennant if either Philadelphia wins their last game or Pittsburgh loses their last game.
   v) If Pittsburgh loses their last game, then Chicago does not win the pennant.
   vi) If either Philadelphia wins their last game and Pittsburgh wins their last game or if Philadelphia loses their last game, then Chicago wins the pennant.
   vii) Chicago wins the pennant if and only if Philadelphia or Pittsburgh loses their last game.

Solution
   i) $(p \wedge \sim q) \vee r$
   ii) $\sim(p \Rightarrow r)$
   iii) $(\sim p \wedge q) \Rightarrow r$
   iv) $(p \vee \sim q) \Rightarrow r$
   v) $\sim q \Rightarrow \sim r$
   vi) $((p \wedge q) \vee \sim p) \Rightarrow r$
   vii) $r \Leftrightarrow (\sim p \vee \sim q)$ ◆

Example 6
Let p represent the algebraic expression, "$x + y < 21$," and q represent, "$m = 65$." Translate each of the following into symbolic language.

i) $x + y < 21$ and $m \neq 65$

ii) $(x + y \geq 21$ and $m = 65)$ or $m \neq 65$

iii) not $(x + y < 21$ or $m = 65)$

iv) $x + y < 21$ or $(x + y \geq 21$ and $m \neq 65)$

Solution

i) $p \wedge \sim q$

ii) $(\sim p \wedge q) \vee \sim q$

iii) $\sim(p \vee q)$

iv) $p \vee (\sim p \wedge \sim q)$ ♦

Exercise Set 3.1

1. Determine which of the following are statements.

    i) Alice owns a restaurant.

    ii) Billy Joel is the greatest rock and roll musician alive.

    iii) Are you going to the dentist on Tuesday?

    iv) $12 + 15 = 30$.

    v) The sun is composed of over 90% hydrogen.

    vi) The meat in the hamburgers at *Burger Universe* tastes like it comes from horses.

2. Determine which of the following are statements.

    i) France is the most beautiful country in Europe.

    ii) France is the largest country in Europe.

    iii) $3x + 5x = 8x$

    iv) Baseball is full of action and excitement.

    v) Presidential elections occur every five years.

    vi) When are you going to visit your parents?

3. Let p represent, "$2x + 40 \leq 50$." Fill in the table below.

x	p	$L(p)$
0		
3		
6		
9		
12		

4. Let p represent, "$r + s = 12$." Fill in the table below.

r	s	p	$L(p)$
−3	28		
−2.5	14.5		
0	12		
4.2	6.8		
6.1	5.9		
20	−8		
22	−12		

5. Let p represent, "$m^2 + n^2 = 100$." Fill in the table below.

m	n	p	$L(p)$
0	10		
3	5		
6	6		
6	8		
7	7		
8	6		
9	1		
10	0		

6.   Let p represent, "$x^2 - 15 \geq 1$." Fill in the table below.

x	p	$L(p)$
1		
2		
3		
4		
5		

7. Let p represent, "I ride my bike everyday."
   Let q represent, "I weigh less than 200 pounds."
Express the following as English statements:

     i) $\sim p$ ii) $p \vee q$ iii) $p \Rightarrow q$ iv) $q \Rightarrow p$

     v) $\sim p \wedge q$ vi) $\sim(p \wedge q)$ vii) $p \Leftrightarrow \sim q$ viii) $\sim(p \Leftrightarrow q)$

8. Let p represent, "Babe Ruth played baseball for the Yankees."
   Let q represent, "The Yankees won the 1927 World Series."
Express the following as English statements:

   i) $p \wedge q$ ii) $p \vee q$ iii) $\sim p$ iv) $p \wedge \sim q$ v) $\sim p \vee (p \wedge q)$

   vi) $p \Rightarrow q$ vii) $\sim q \Rightarrow \sim p$ viii) $p \Leftrightarrow q$ ix) $\sim p \vee q$ x) $\sim(p \Leftrightarrow q)$

9. Let p represent, "It is summer on Mars."
Let q represent, "Dust storms are occurring on Mars."
Let r represent, "The brightness of Mars changes."
Translate each of the following into symbolic language.

www.shutterstock.com · 4814683

   i) The brightness of Mars changes whenever dust storms are occurring.
   ii) Dust storms are occurring on Mars and it is not summer on Mars, but its brightness is not changing.
   iii) It is neither summer on Mars nor are dust storms occurring.
   iv) Dust storms are not occurring on Mars whenever its brightness is not changing.
   v) If it is not summer on Mars, then either dust storms are occurring or its brightness is changing.
   vi) It is summer on Mars if and only if dust storms are not occurring.
   vii) The brightness of Mars changes if and only if it is not summer and dust storms are occurring.
   viii) It is not the case that either it is summer on Mars or dust storms are occurring while its brightness changes.

10.   Let p represent, "Scott plays soccer."

Let q represent, "Scott plays the violin."

Let r represent, "Scott does well in school."

Translate each of the following into symbolic language.

   i) Scott plays soccer, but not the violin.

   ii) Scott either does not play soccer or he does not play the violin.

   iii) Scott plays either soccer or the violin, but he does not do well in school.

   iv) Scott plays soccer if and only if he does not play the violin.

   v) If Scott plays soccer, he does well in school.

   vi) Whenever Scott does well in school, he plays either soccer or the violin.

   vii) Scott plays the violin but not soccer whenever he does not do well in school.

   viii) If Scott either does well in school or plays the violin, then he plays soccer.

11. Identify the antecedent and the consequent in the following conditional statements.

   i) All square polygons are rectangles.

   ii) Every rational number is a real number.

   iii) A magnetic field is created whenever an electric current flows through a wire.

   iv) His movies make a profit whenever they gross \$50 million or are seen by 10 million people.

   v) Much debate occurs in Congress every time the topic is gun control or welfare.

   vi) We go fishing on the weekend provided that it doesn't rain or snow.

12. Identify the antecedent and the consequent in the following conditional statements.

   i) Whenever it rains, it pours.

   ii) All humans are mammals.

   iii) If you plunge into the stock market, you either sink or swim.

   iv) Neither Ashley nor Betty go to the beach if the temperature is not above 75°.

   v) You receive a driver's license provided you are sixteen years old and you pass the exam.

   vi) All minerals that are diamonds are composed of carbon.

13. Let p represent, "$x \geq 7$."

   Let q represent, "$5x < y$."

   Let r represent, "$y = 350$."

Translate each of the following into symbolic language.

   i) $x \geq 7$ or $y \neq 350$

   ii) $(x < 7$ and $5x < y)$ or $y = 350$

   iii) $x \geq 7$ and $5x \geq y$ and $y \neq 350$

   iv) $x < 7$ or not $(5x < y$ and $y = 350)$

   v) $(x \geq 7$ and $5x < y)$ or $(5x \geq y$ and $y = 350)$

14. Let p represent, "$x + y = 2500$."

   Let q represent, "$8w > z$."

   Let r represent, "$2x + 5y - 11z \leq w$."

Translate each of the following into symbolic language.

i) $x + y = 2500$ or $8w \leq z$

ii) $(x + y \neq 2500$ and $2x + 5y - 11z \leq w)$ or $8w > z$

iii) not $(8w > z$ or $2x + 5y - 11z > w)$ and $x + y = 2500$

iv) $8w \leq z$ and not $(2x + 5y - 11z \leq w$ and $x + y = 2500)$

v) not $((x + y \neq 2500$ and $8w > z)$ or $(8w > z$ and $2x + 5y - 11z > w))$

15. A *set* is defined as any collection of objects, regardless of whether those objects are numbers, cars, galaxies, or hairs on your head. Bertrand Russell once defined a specific set R whose objects were themselves sets by

$$R = \{ S \mid S \text{ is a set which is not a member of itself.}\}$$

Is the sentence, "R is a member of R," a statement? (This example is known as *Russell's Paradox*.)

16. Scott makes the statement, "I am the Diabolical Disagreer. I disagree with all claims made by anyone." Nick says to Scott, "I claim you disagree with me." Is this a paradox?

17. Korlin lives in a home with ten other people of all ages. Each person has a special job and Korlin's job is to shine *only* the shoes of every person who does not shine his or her own shoes. The question is: Does Korlin shine her own shoes? Is this a paradox?

3.2 Truth and Consequences

Every compound statement has a logical value which is a function of the values of its prime statements. For instance, if you are currently attending college, then the statement,

p: I am a college student.

is true. On the other hand,

q: I am an ax murderer.

is (hopefully) false. You would agree then that the compound statement, "I am a college student and an ax murderer," is also false. In other words, any pair of statements p, q which are joined by the conjunction "and" form a false compound statement unless both p is true and q is true. Using the logical function L, we say that the only time that $L(p \wedge q)$ is "True" is when $L(p)$ is "True" and $L(q)$ is "True".

A chart known as a ***truth table*** places "T" for the value "True" and "F" for the value "False" in every box to indicate the appropriate logical value for a compound statement given the various possible combinations for the prime statements. The truth table for conjunction is given below. Although the columns should be headed by $L(p)$, $L(q)$, and $L(p \wedge q)$, we just use p, q, and $p \wedge q$ for the sake of convenience. This convention will be followed in all ensuing truth tables.

p	q	p ∧ q
T	T	T
T	F	F
F	T	F
F	F	F

Figure 3.2.1 Truth table for conjunction.

Example 1
Determine the truth of the statement, "George Washington was the first US president and Abraham Lincoln was the second."

Solution
If p represents, "George Washington was the first US president," and q represents, "Abraham Lincoln was the second US president," then p is true, q is false, and the given statement – represented by $p \wedge q$ – is false. ◆

Example 2
Under what conditions would the statement, "Deep space is cold and dark," be true?

Solution
Both the prime statements, "Deep space is cold," and "Deep space is dark," must be true. ◆

Note that the table in Figure 3.2.1 does not depend in any way on the statements represented by p and q. This is the essential feature of classical **symbolic logic**, also called logical calculus, which uses formal abstractions in order to systematize and codify principles of valid reasoning. We are not concerned with the truth of the prime statements in a given argument,

but rather whether the argument itself is valid. One of the pioneers of this endeavor was the famous British mathematician **George Boole** (1815–1864). Born into extreme poverty, Boole struggled valiantly as a child to raise his standing in society by learning Latin and Greek on his own. His success at translating the classics led to a job as an elementary school teacher while yet a teenager which, in turn, inspired him to study higher mathematics. While only 20, his natural gift for the discipline blossomed forth as he took up the study of the great works of Lagrange and Laplace, the successors to Isaac Newton. In 1848, he published *The Mathematical Analysis of Logic*, in which he wrote,

They who are acquainted with the present state of the processes of analysis does not depend upon the inte̶ solely on the laws of their combination.

This book was the precursor to Boole's masterpiece, *The Laws of Thought*, about which Bertrand Russell said, "Pure Mathematics was discovered by Boole in [this] work." The use of truth tables for the logical examination of argument began in part from this highly original book.

The tables for negation and disjunction are given in Figure 3.2.3.

p	~p
T	F
F	T

p	q	p ∨ q
T	T	T
T	F	T
F	T	T
F	F	F

Figure 3.2.3 Truth tables for negation and disjunction.

The negation of a true statement is false and the negation of a false statement is true. For instance, if "Harry is a swimmer" is false, then "Harry is a non-swimmer" (or "Harry is not a swimmer," or "Harry does not swim") is true.

In everyday English, one often uses "or" in the exclusive sense such as in, "I am going to the opera or I am going to the concert," where the speaker is usually implying that he will not attend both. However, in symbolic logic the "or" used in the disjunction of two statements is always considered to be *inclusive*. Therefore it is always a possibility that both statements p and q can be true and in that case we see in Figure 3.2.3 that $p \vee q$ is true.

Example 3

Determine the logical value of, "Earth's surface consists of more water than land and Elvis Presley is alive." Determine the logical value of, "Earth's surface consists of more water than land or Elvis Presley is alive."

Solution

Let p represent the first prime statement and q the second. Since p is true and q is false, the first compound statement $p \wedge q$ is false and the second compound statement $p \vee q$ is true. ♦

Example 4

Consider the algebraic statements, p: "$x \geq 13$" and q: "$x \leq 29$," where x is a real number variable. Since p and q are expressions involving x, their values change as x changes and so we use the L function to indicate their current logical value.

x	p	q	$L(p)$	$L(q)$	$L(p \wedge q)$	$L(p \vee q)$
10	$10 \geq 13$	$10 \leq 29$	F	T	F	T
13	$13 \geq 13$	$13 \leq 29$	T	T	T	T
20	$20 \geq 13$	$20 \leq 29$	T	T	T	T
30	$30 \geq 13$	$30 \leq 29$	T	F	F	T ♦

We now examine the possible logical configurations for the conditional statement,

"If Debra does your taxes, then your taxes are done correctly."

If both the antecedent, "Debra does do your taxes," and the consequent, "Your taxes are done correctly," are true, then certainly the entire conditional is true. Furthermore, if Debra does not do your taxes then the fact of whether or not they are done correctly does not affect the truth of the conditional and so it is true regardless of the outcome. It is only if Debra does your taxes and they are done incorrectly when we see that the original conditional is false. Figure 3.2.4 displays the truth table for the conditional.

p	q	p $\Rightarrow$ q
T	T	T
T	F	F
F	T	T
F	F	T

Figure 3.2.4 Truth table for the conditional.

Example 5

93

We see that some rather bizarre conditional statements can be true. "If it rains milk, then Zebras have polka dots," is a true statement since it is impossible for the antecedent to be true. The same can be said for, "If $6 + 6 = 13$, then everyday is a holiday." It is just as important to realize that a true consequent guarantees a true conditional. "If your middle name is *Bozo*, then Harrisburg is the capital of Pennsylvania," is always true. ♦

To see the difference between a conditional and a biconditional, suppose the following two statements are *true*.

I: "If Jack receives a flu shot in October, then he does not get the flu in winter."
II: "Jack does not get the flu in winter if and only if he receives a flu shot in October."

The conditional statement I tells us that a true antecedent – Jack receives a flu shot in October – means that the consequent must also be true – Jack does not come down with the flu in winter. However, we know that a false antecedent does not guarantee a false consequent even though statement I is true. Jack may get lucky and not get the flu even if does not get the appropriate shot.

A true biconditional statement II is stronger than statement I. It states that Jack does not get the flu in winter if and *only if* he has a flu shot in October. This gives more information than statement I by stating that when, "he receives a flu shot in October," is false, then, "Jack does not get the flu," is also false. In other words, Jack's immune system is such that the only way he avoids the flu is when he receives the shot. Either both prime statements are true or both are false. *In a true biconditional the logical values of the prime statements are the same.* Similarly, it follows that if the logical values of the two primes are different, then the resulting biconditional is false. These facts are summarized in Figure 3.2.5.

p	q	$p \Leftrightarrow q$
T	T	T
T	F	F
F	T	F
F	F	T

Figure 3.2.5 Truth table for the biconditional.

Example 6

Determine the logical value of, "The water in the pond turns into ice if and only the temperature is above 32° F."

Solution

We know this to be a false statement from common experience for if the first prime statement is false, then the second one is true. Also, if the first prime is true, then the second is false. Since the only situation that exists is

94

for the p and q of this biconditional statement to have opposite logical values, the value of the biconditional is false. ♦

We now return to our characterization of symbol logic as *objects* and *operations*. The statement symbols are the objects and the tables given in the above figures can be considered to be definitions of the operations $\sim$, $\wedge$, $\vee$, $\Rightarrow$, and $\Leftrightarrow$ on those objects. With these definitions we can assign a logical value (T or F) to *any* expression of objects and operations without regard or need for the objects to have an English meaning, but based solely on the values of the prime component symbols. Such expressions – having logical value but no meaning – are called *statement forms*. Any substitution of English words for the object and operation symbols in a form is then called an **interpretation** of the form. (Hence, all of the translations in the last section were back and forth from statement forms to interpretations.) For simplicity, we will continue to call the values T "true" and F "false" when referring to the values of statement symbols and forms. The standard order in which operations are performed is: $\sim$, $\wedge$, $\vee$, $\Rightarrow$, and $\Leftrightarrow$ except where parentheses are used. As in algebra, any operation within a pair of parentheses is to be done first. Even when they do not alter the standard order, parentheses are often used for clarification and, in the case of nested parentheses, you begin with the operation inside the innermost pair.

Example 7

Let p, q, r, and s be statement symbols having values T, F, T, and F respectively. Find the values of the following statement forms.

    i) $\sim p \wedge q \vee r$
    ii) $\sim p \wedge (q \vee r)$
    iii) $\sim p \wedge q \Rightarrow r \wedge \sim s$
    iv) $\sim(p \wedge q \vee r) \Rightarrow r \wedge s$
    v) $(p \Rightarrow q) \Leftrightarrow (r \Rightarrow s)$
    vi) $\sim(p \wedge q \Leftrightarrow r) \wedge (\sim r \Rightarrow s)$

Solution

i) Negation is done first followed by conjunction and then disjunction.

$\sim p$	$\wedge$	q	$\vee$	r	
F		F		T	$\sim p$ has value F
	F			T	$\sim p \wedge q$ has value F
			T		$\sim p \wedge q \vee r$ has value T

ii) This time we do disjunction after negation because of the parentheses.

$\sim p$	$\wedge$	$(q$	$\vee$	$r)$	
F		F		T	$\sim p$ has value F
F			T		$q \vee r$ has value T
	F				$\sim p \wedge (q \vee r)$ has value F

iii)
```
iii)   ~p   ∧   q   ⇒   r   ∧   ~s
        F        F        T        T
            F                 T
                     T
```

iv)
```
iv)   ~(p   ∧   q   ∨   r)   ⇒   r   ∧   s
        T        F        T            T        F
            F             T                     F
                      T                         F
                      F
                           T
```

v)
```
v)    (p   ⇒   q)   ⇔   (r   ⇒   s)
       T        F         T        F
           F                  F
                   T
```

vi)
```
vi)   ~(p   ∧   q   ⇔   r))   ∧   (~r   ⇒   s)
        T        F        T             F        F
            F             T                  T
                    F                        T
                    T                        T
                         T                              ♦
```

The truth table for any statement form must contain a value for every combination of values for the prime statements. Consider the form $p \wedge \sim q$. In Figure 3.2.6 you see that first the values for p are repeated and $\sim q$ is evaluated. Then the values for the entire form are computed in the boxed column headed by the $\wedge$ by operating on each pair of the listed values for p and $\sim q$.

p	q	p ∧ ~q	
T	T	T	F
T	F	T	T
F	T	F	F
F	F	F	T

p	q	p	∧	~q
T	T	T	**F**	F
T	F	T	**T**	T
F	T	F	**F**	F
F	F	F	**F**	T

Figure 3.2.6 Truth table for $p \wedge \sim q$.

Example 8

Consider the longer form: $(p \vee \sim q) \Rightarrow \sim(p \wedge q)$. The order of operations is important and is indicated by the numbering at the bottom of the columns in the sequence of tables in Figure 3.2.7. The first step is to list the values for p, q, and $\sim q$. Second, we determine the values for each connective inside the parentheses. Third, we reverse the values of $(p \wedge q)$ because of the " $\sim$ " symbol. The fourth and final boxed column of bold-face values contains the logical values resulting from connecting the values in the oval enclosed columns using the conditional ($\Rightarrow$).

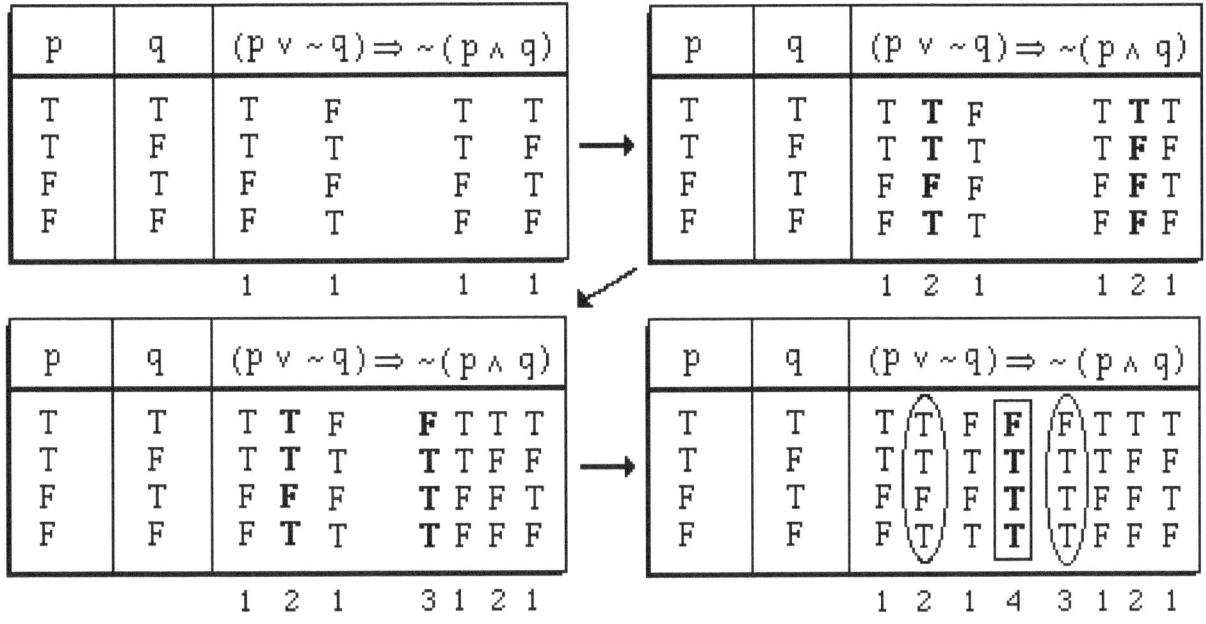

Figure 3.2.7 The truth table for $(p \lor \sim q) \Rightarrow \sim(p \land q)$.

Example 9

We now evaluate a truth table for a form containing three statement symbols p, q, and r. Since each of three symbols can be one of two logical values, we must examine $2^3 = 8$ combinations of these values as shown in the table below. In general, one must consider 2^n combinations for a statement form composed of n statement symbols. In Figure 3.2.8 we show the results for the form: $(p \lor q) \land r$. Disjunction ($\lor$) is computed before conjunction ($\land$) because of the parentheses. ♦

p	q	r	(p	∨	q)	∧	r
T	T	T	T	T	T	**T**	T
T	T	F	T	T	T	**F**	F
T	F	T	T	T	F	**T**	T
T	F	F	T	T	F	**F**	F
F	T	T	F	T	T	**T**	T
F	T	F	F	T	T	**F**	F
F	F	T	F	F	F	**F**	T
F	F	F	F	F	F	**F**	F
			1	2	1	3	1

Figure 3.2.8 Truth table for a form with three prime statement symbols.

Sometimes you can deduce the logical value of a given statement form if you know the value of a related form.

Example 10

If the value of $p \Leftrightarrow q$ is true, what is the value of $p \Rightarrow q$?

97

Solution

From Figure 3.2.5, we see that if $p \Leftrightarrow q$ is true, then p and q are either both true or both false. In either case, the value of $p \Rightarrow q$ must be true.

Note that $p \Rightarrow q$ does not have a unique value if $p \Leftrightarrow q$ is given to be false. In that event, either p is true and q is false (in which case $p \Rightarrow q$ is false) or p is false and q is true (in which case $p \Rightarrow q$ is true). ◆

Example 11

Based on the given information, determine whether the statement form has a unique logical value and, if so, find it.

i) $(p \wedge \sim q) \Rightarrow (r \vee \sim s)$
    T

ii) $p \wedge (q \Leftrightarrow r)$
    T

iii) $(p \wedge q) \Leftrightarrow (\sim p \vee r)$
    F

Solution

i) Since the value of r is T, the value of $r \vee \sim s$ must be T. The value of the conditional must then be T, because the consequent has a value of T.

ii) Knowing the value of the biconditional does not give a unique value to the entire conjunction. If p is true, then $p \wedge (q \Leftrightarrow r)$ is true. If p is false, then $p \wedge (q \Leftrightarrow r)$ is false.

iii) p has value F and so $\sim p$ has value T. Therefore $p \wedge q$ has value F, $\sim p \vee r$ has value T, and $(p \wedge q) \Leftrightarrow (\sim p \vee r)$ has value F. ◆

Suppose a friend on the phone tells you it is hot and sunny outside, but you look out the window and you see that, even though the temperature is 98°, it is cloudy . So you disagree with your friend based on the lone fact that it is cloudy. In other words, you negated the entire compound statement, "It is hot and sunny," because the negation conveys the same meaning as, "It is either not hot or it is not sunny." It often happens that two different statements have the exact same meaning. Formally we say that two statement forms are **logically equivalent** or just **equivalent** if their truth tables have identical final columns. We denote the phrase "is equivalent to" by the symbol ≡ . For instance, it is clear that $\sim(\sim p) \equiv p$. Also, we have just seen an interpretation of the equivalence

$$\sim(p \wedge q) \equiv \sim p \vee \sim q.$$

where p is, "It is hot," and q is, "It is sunny." This is verified by the following truth table.

p	q	~ (p ∧ q)	~ p ∨ ~ q
T	T	**F** T T T	F **F** F
T	F	**T** T F F	F **T** T
F	T	**T** F F T	T **T** F
F	F	**T** F F F	T **T** T

98

Figure 3.2.9 Equivalent statement forms.

For simplicity, we also refer to interpretations of equivalent forms as equivalent interpretations.

Example 12

Computer scientists utilize the equivalence in Figure 3.2.9 to implement negations of compound algebraic statements in computer programs. Find equivalent statements to the negation of each of the following statements.

i) $x = 18.5$ and $y > 31.5$
ii) $x \geq 70$ and $x \leq 75$
iii) $a^2 \neq 1000$ and $a + b < 250$

Solution

i) $x \neq 18.5$ or $y \leq 31.5$
ii) $x < 70$ or $x > 75$
iii) $a^2 = 1000$ or $a + b \geq 250$ ◆

Example 13

"If Marlowe located Jessica Vanderkamp, then he earned \$5000," is an interpretation of the conditional $p \Rightarrow q$. Show that this form is equivalent to $\sim q \Rightarrow \sim p$ and write out the corresponding interpretation.

Solution

The equivalence is verified by this truth table.

p	q	p $\Rightarrow$ q			$\sim q \Rightarrow \sim p$		
T	T	T	**T**	T	F	**T**	F
T	F	T	**F**	F	T	**F**	F
F	T	F	**T**	T	F	**T**	T
F	F	F	**T**	F	T	**T**	T

Figure 3.2.10 Equivalent statement forms.

The corresponding interpretation is, "If Marlowe did not earn \$5000, then he did not locate Jessica Vanderkamp."   ◆

In our study of Galileo and his rationale for supporting the Copernican system (Section 3.3), it was stated that $\sim q \Rightarrow \sim p$ is known as the ***contrapositive*** of the conditional $p \Rightarrow q$. Figure 3.2.10 demonstrates that these are logically equivalent statements.

On the other hand the statement, "If Marlowe earned \$5000, then he located Jessica Vanderkamp," in the above example would be an interpretation of $q \Rightarrow p$, which does *not* have the same final truth values as $p \Rightarrow q$ and thus is *not* equivalent to it. In the Section 3.3 it was also stated that these related conditional forms, $p \Rightarrow q$ and $q \Rightarrow p$, are called **converses** of each other.

Example 14

      Determine the converse and contrapositive of the conditional statement, "If $f(x) = 4x + 10$, then $f(2) = 18$."

Solution

      The converse is obtained by simply exchanging the antecedent and consequent. In the present example this is, "If $f(2) = 18$, then $f(x) = 4x + 10$." We know that many functions exist for which $f(2) = 18$ such as $f(x) = 5x + 8$ or $f(x) = 12x - 6$. Therefore this converse is not a true statement. However, the contrapositive is, "If $f(2) \neq 18$, then $f(x) \neq 4x + 10$." This is certainly true and exemplifies the fact that any conditional statement is logically equivalent to its contrapositive. ◆

      Translating two given compound statements into forms and then comparing truth tables is an excellent method for determining equivalence. This strikes at the heart of the social and scientific utility of logical abstraction of complex statements. The symbols we use for statements are nothing more than indifferent variables whose combinations (forms) have values determined according to our defined set of operations. The replacement of English by logical symbolism allows for dispassionate comparison of seemingly similar sentences and, moreover, effective analysis of long involved proofs of mathematical facts as well as philosophical or political *argument*. This will be explored more thoroughly in the next section.

Exercise Set 3.2

1. Under what conditions would the statement, "Nick enjoys baseball and spinach," be true?

2. Under what conditions would the statement, "Andre lives in Spain and Asif lives in Pakistan," be false?

3. Determine the logical value of, "$289 > 147$ and $35 < 11$." Determine the logical value of "$289 > 147$ or $35 < 11$."

4. Determine the logical value of, "The US Senate consists of 100 senators or the House of Representatives consists of 100 representatives." Determine the logical value of, "The US Senate consists of 100 senators and the House of Representatives consists of 100 representatives."

5. Determine the logical value of, "If $6 + 8 = 10$, then the Earth is made of sushi." Determine the logical value of, "If $6 + 8 \neq 10$, then the Earth is made of sushi."

6. Determine the logical value of, "Your math teacher can lift 500 pounds whenever your teacher can run 100 meters in 3.5 seconds." Determine the logical value of, "Your math teacher can lift 500 pounds whenever your teacher can run 100 meters in less than 10 minutes."

7. Let p represent, "$x > 1000$," and let q represent, "$x < 1500$." Fill in the following table.

x	p	q	$L(p)$	$L(q)$	$L(p \wedge q)$	$L(p \vee q)$
500						
1000						
1250						
1499						
1500						
1600						

8. Let p represent, "$x > 0$," and let q represent, "$2x + y \leq 26.3$." Fill in the following table.

x	y	p	q	$L(p)$	$L(q)$	$L(p \wedge q)$	$L(p \vee q)$
−5	45						
−1	28						
2	20						
10	5						
12	3						

9. Let p represent, "$x \geq 75$," and let q represent, "$x + y = 100$." Fill in the following table.

x	y	p	q	$L(p)$	$L(q)$	$L(p \wedge q)$	$L(p \vee q)$
62	15						
70	30						
75	25						
80	20						
90	11						

Let the statement symbols p, q, r, and s have values T, F, T, *and* F *respectively. Find the values of the following statement forms.*

10. $p \land q \lor r \land s$

11. $(p \lor q \land r) \lor s$

12. $\sim(p \land r) \Rightarrow q \lor s$

13. $(q \Leftrightarrow s) \Rightarrow (\sim r \Leftrightarrow \sim s)$

14. $((p \lor r) \land s) \Leftrightarrow (q \lor r)$

15. $((q \Leftrightarrow s) \land p) \land \sim(p \land r \Rightarrow s)$

Evaluate the truth table for each of the following statement forms.

16. $p \lor \sim q$

17. $\sim p \Rightarrow q$

18. $\sim p \land q$

19. $\sim(p \land q)$

20. $\sim q \Rightarrow \sim p$

21. $p \land \sim p$

22. $(p \land q) \Rightarrow \sim(p \lor q)$

23. $(p \Rightarrow q) \land (q \Rightarrow p)$

24. $(p \land \sim q) \Rightarrow \sim r$

25. $(p \land q) \lor r$

26. $(p \Rightarrow q) \land (p \Rightarrow \sim q)$

27. $p \Rightarrow (q \lor \sim r)$

Based on the given information, determine whether the statement form has a unique logical value and, if so, find it.

28. $p \Rightarrow (q \lor \sim r)$
    F

29. $p \lor (q \Rightarrow r)$
    T

30. $(p \lor q) \Leftrightarrow (r \land s)$
  T   F

31. $(p \lor q) \Leftrightarrow (r \land s)$
  F   T

32. $(p \Rightarrow q) \Rightarrow (\sim q \Rightarrow s)$
    T

33. $(p \land \sim p) \Leftrightarrow (q \lor \sim q)$

34. If the value of $p \wedge q$ is F, what is the value of $\sim p \vee \sim q$?

35. If the value of $p \vee q$ is F, what is the value of $p \Leftrightarrow q$?

36. If the value of $p \Leftrightarrow q$ is T, what is the value of $p \Leftrightarrow \sim q$?

37. If the value of $p \Rightarrow q$ is T, what is the value of $p \wedge \sim q$?

Verify the following equivalences.
38. $p \Leftrightarrow q \equiv \sim p \Leftrightarrow \sim q$ **39.** $p \Leftrightarrow q \equiv (p \Rightarrow q) \wedge (q \Rightarrow p)$

40. $\sim(p \Rightarrow q) \equiv p \wedge \sim q$ **41.** $p \Rightarrow q \equiv \sim p \vee q$

42. Two pairs of equivalent forms known as *De Morgan's Laws* are:
$$\sim(p \wedge q) \equiv \sim p \vee \sim q$$
$$\sim(p \vee q) \equiv \sim p \wedge \sim q$$
The first equivalence was verified in Figure 3.2.9. Use a truth table to verify the second equivalence.

43. Use De Morgan's Laws to find a statement equivalent to, "It is not the case that Senator Farnsworth voted for the tax cut and for the new spending bill."

44. Use De Morgan's Laws to find a statement equivalent to, "Last year, average family income did not increase and average family spending did not increase."

45. Use De Morgan's Laws to find a statement equivalent to, "It is not the case that Erika received either a B in Math or a C in English."

46. Use De Morgan's Laws to find a statement equivalent to, "In the soccer game, either Jorge scored a goal or he made an assist."

Use De Morgan's Laws to find the negation of each of the following algebraic statements.
47. $x > 2$ and $y = 35$ **48.** $x \neq 60$ or $x + y \leq 320$

49. $x + y = 4$ or $x < 13.5$ **50.** $x \geq 56$ and $x \leq 57$

51. $y < 85$ or $y > 90$ **52.** $r > 6$ and $r^2 + s^2 = 100$

53. $(x < 10$ and $y < 10)$ or $2x + 3y = 50$ **54.** $(a = 25$ or $b = 25)$ and $a + b > 600$

55. Determine the converse and the contrapositive of the conditional statement, "If the earth's core is made principally of iron, then the temperature of the core is between 4000 and 5000° C." Which one is equivalent to the original?

56. Determine the converse and the contrapositive of the conditional statement, "If $m = 8$, then $m^2 = 64$." Which one is equivalent to the original?

57. Determine the converse and the contrapositive of the conditional statement, "If $f(x) = 8x + 3$, then $f(2) = 19$." Which one is equivalent to the original?

58. Determine the converse and the contrapositive of the conditional statement, "If the length and width of a rectangle are 5 cm and 10 cm, then its area is 50 cm^2." Which one is equivalent to the original?

*59. Suppose we are only allowed to use the connectives $\sim$ and $\wedge$. We could implement the other connectives by use of equivalences. For instance, we could write $\sim(\sim p \wedge \sim q)$ for $p \vee q$ according to DeMorgan's Laws. What would you write for $p \Rightarrow q$ using only $\sim$ and $\wedge$? [*Hint*: See #40.]

*60. What would you write for $p \Leftrightarrow q$ using only $\sim$ and $\wedge$? [*Hint*: See #39.]

*61. In computer science programming, the function $\mathbf{NOR}(p, q)$ assumes the logical values defined by $\mathbf{NOR}(p, q) \equiv \sim p \vee \sim q$. All the connectives can be implemented via this one function (called a logic gate) by varying the arguments of the function. For instance, $\sim p \equiv \mathbf{NOR}(p, p)$ and so $p \vee q \equiv \mathbf{NOR}(\mathbf{NOR}(p, p), \mathbf{NOR}(q, q))$. How could you express $p \wedge q$ using $\mathbf{NOR}$?

*62. Another special function used by computer programmers is $\mathbf{XOR}$, the "exclusive or". $\mathbf{XOR}(p, q)$ assumes the value true if either p is true (and q is false) or q is true (and p is false), but is false in the case that p and q have the same logical value. In other words, it is the same as $p \vee q$ except when p and q are both true. Create a statement form that is equivalent to this function.

3.3 Tautologies and Syllogisms

God has not been so sparing of men to make them barely two-legged creatures, and left it to Aristotle to make them rational.

John Locke

I thought of mathematics with reverence, and suffered when Wittgenstein led me to regard it as nothing but tautologies.

Bertrand Russell

We discussed the impact of Bertrand Russell on modern thought in the first section, but we should mention that, during his lifetime, his views were not always held in high esteem by everyone. In particular, he was an outspoken peace activist and his support of the No-Conscription Fellowship during World War I embroiled him in controversy. Apparently, you can get yourself into no small amount of trouble by being too logical.

> According to cable reports from London, the Council of Trinity College, Cambridge, has removed Professor Bertrand Russell from his lectureship in logic and principles of mathematics on account of his having been convicted under the defense of the realm act for publishing a leaflet defending the "Conscientious Objector" to service in the British army.
>
> *American Mathematical Monthly*
> Vol. 23, 1916, p. 317

In fact, Russell spent several productive months in jail in 1918 on charges resulting from the authorship of inflammatory anti-war literature. Prison, however, seemed not to dampen his creative spirit for he wrote much of his *Introduction to Mathematical Philosophy* (from which a quote appears at the beginning of this chapter) during this time. This work contained certain results stemming from the framework for symbolic logic constructed by George Boole and outlined the methodology of the theory of deduction. **Deduction** is the process of reaching a **conclusion** from a given set of **premises**. A *premise* may be a prime statement or a compound statement and an entire sequence of premises leading to a conclusion is called an **argument**. We are primarily concerned here with the determination of the *validity* of an argument. Simply stated, an argument is considered to be **valid** if it is impossible for the conclusion to be false when the premises are true. The use of a valid argument to obtain a conclusion is called *valid reasoning*. It is important to realize that this does *not* state that the validity of an argument is a property equivalent to the truth of its conclusion. Before we introduce a process for determining validity, we define an associated characteristic of statements in general.

Consider the statement form $(p \wedge q) \Rightarrow p$ and its associated truth table.

p	q	(p ∧ q) ⇒ p
T	T	T T T **T** T
T	F	T F F **T** T
F	T	F F T **T** F
F	F	F F F **T** F

Figure 3.3.1 Example of a tautology.

Note that the logical value of this compound statement is true under all possible combinations of values for the prime statements. Such a statement form is called a ***tautology***. Realize that tautologies exist independently of the truth of the prime statements. They simply guarantee that the statement taken *as a whole* is always true. Four interpretations of the above form which represent each row of the table in Figure 3.3.1. are:

If horses are mammals and sharks are predators, then horses are mammals.

If horses are mammals and Ben Franklin is alive, then horses are mammals.

If horses are reptiles and sharks are predators, then horses are reptiles.

If horses are reptiles and Ben Franklin is alive, then horses are reptiles.

All of these interpretations are true statements.

Example 1

Determine whether the following statement forms are tautologies.

(i) $(p \Leftrightarrow \sim q) \wedge p \Rightarrow \sim q$

(ii) $p \wedge (q \vee r) \Rightarrow p \wedge r$

Solution

(i)

p	q	(p ⇔ ~q) ∧ p ⇒ ~q
T	T	T F F F T **T** F
T	F	T T T T T **T** T
F	T	F T F F F **T** F
F	F	F F T F F **T** T

          1 2 1 3 1 4 1

The parentheses dictate that we evaluate the biconditional first. By order of operations we evaluate conjunction next followed by the conditional. Since all the final values are true, this statement form is a tautology.

(ii)

106

p	q	r	p	∧	(q	∨	r)	⇒	p	∧	r
T	T	T	T	T	T	T	T	**T**	T	T	T
T	T	F	T	T	T	T	F	**F**	T	F	F
T	F	T	T	T	F	T	T	**T**	T	T	T
T	F	F	T	F	F	F	F	**T**	T	F	F
F	T	T	F	F	T	T	T	**T**	F	F	T
F	T	F	F	F	T	T	F	**T**	F	F	F
F	F	T	F	F	F	T	T	**T**	F	F	T
F	F	F	F	F	F	F	F	**T**	F	F	F

1 3 1 2 1 4 1 3 1

The parentheses dictate that we evaluate the disjunction before the conjunctions. The conditional (⇒) is used to operate on pairs of values occurring in the two columns under the conjunctions. The occurrence of a false in the second row of the final column prevents this statement form from being a tautology. ♦

When programming a computer, algebraic statements are used to provide direction for the flow of the execution of the program. The purpose of such a statement might be considered to be the control of traffic at a two-pronged fork in the road. Program execution continues along one path if the value of the expression is true and along the other path if the value of the expression is false. So the expression serves as a sort of traffic light. An algebraic statement which has a constant logical value for all possible values of the variables channels the flow into just one branch and essentially blocks off the remaining branch. Although not a tautology in the strict sense of the word, it is a related type of phenomenon in computer programming creating a type of mistake called a *logic error*, since it is certainly not what the programmer intended. It serves no purpose – analogous to putting in a traffic light at an intersection that always remains green!

Example 2

Although "$x + y \geq 30$" could be true or false depending on the values of x and y, the compound statement, "$x + y \geq 30$ or $x^2 \geq 0$" will *always* be true, because the square of any number is never less than zero. Since "$x^2 \geq 0$" is always true, so is "$x + y \geq 30$ or $x^2 \geq 0$." Figure 3.3.2 displays the two situations with a device known in computer science as a *flow chart*. Note that the algebraic expression is written inside of a diamond.

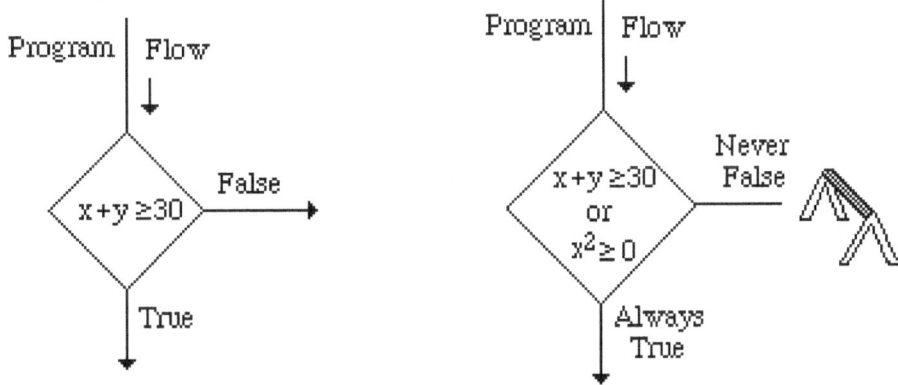

Figure 3.3.2 The perpetually true algebraic expression on the right is a programming error.

Other examples are:

$$x > 25 \text{ or } x \le 25 \qquad\qquad 4x + 9y = 17 \text{ or } 4x + 7y \ne 17$$

Similarly, a logic error occurs when an algebraic statement is always false. Examples are:

$$2x + y \ge 100 \text{ and } 2x + y < 100 \qquad\qquad (x > 3 \text{ and } y > 3) \text{ and } x + y = 2$$

Extreme care must be taken to avoid these types of mistakes when writing computer programs. ♦

We first encountered an argument known as a *syllogism* in Section 2.3 in connection with the logic used by **Galileo Galilei** (1564–1642) to reject the Ptolemaic arrangement of the planets.

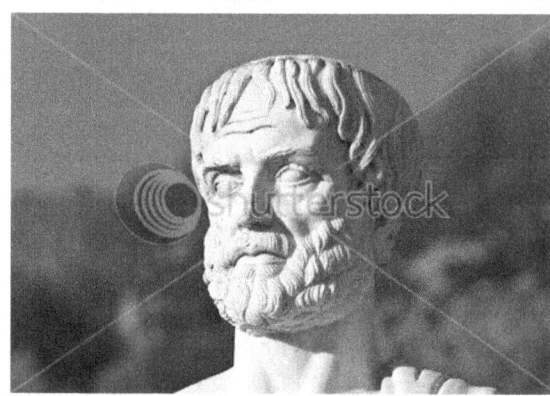

Officially, this is one type of an argument consisting of two premises and a conclusion, each pair of which have a connection such as a common object, description, or action. The term first derived from **Aristotle** (384–322 B.C.) who also was previously mentioned in Chapter 2. Aristotle initiated the systemization of logical argument by categorizing the various types of syllogisms. One common type is exemplified by:

Figure 3.3.3 Statue of Aristotle (384–322 BC) in his birthplace Stageira, Greece

Premise: If consumer prices increase, then purchasing power decreases.

Premise: Consumer prices are increasing.

Conclusion: ∴ Purchasing power is decreasing.

Syllogisms are often structured according to the above format. The two premises are listed above a horizontal line with the conclusion stated below it. The symbol "∴" means "therefore".

For an accurate logical analysis, we first translate this argument into symbolic language. Let p represent, "consumer prices increase," and let q represent, "purchasing power decreases." Then we can join the premises of the above syllogism with a conjunction to create the compound statement, $(p \Rightarrow q) \wedge p$. (Liberties may be taken with the expression and tense of the verbs.) The conclusion of this argument is then written as the consequent of the conditional statement form,

$$[(p \Rightarrow q) \wedge p] \Rightarrow q$$

which we see in Figure **3.3**.4 is a tautology.

p	q	[(p	⇒	q)	∧	p]	⇒	q
T	T	T	T	T	T	T	**T**	T
T	F	T	F	F	F	T	**T**	F
F	T	F	T	T	F	F	**T**	T
F	F	F	T	F	F	F	**T**	F

$$1 \quad 2 \quad 1 \quad 3 \quad 1 \quad 4 \quad 1$$

Figure 3.3.4 This syllogism is a valid argument known as *modus ponens*.

This means that it is impossible for the conclusion to be false if the premises are true, a fact which is incorporated into our definition of the validity of any argument.

> *Whenever the truth table of the corresponding statement form of an argument is a tautology, the argument is said to be* **valid**.

So we see that any syllogism having the form $[(p \Rightarrow q) \wedge p] \Rightarrow q$ is a valid argument. Since this particular reasoning scheme is so common, it has a special nomenclature. It is known both as *direct reasoning* and also by the label ***modus ponens***, which we will abbreviate by *MP*.

Another common type of syllogism which falls in the category of *indirect reasoning* can be illustrated by a review of Aristotle's "proof" that the Earth is a sphere. (See Section 3.2.) Recall that Aristotle rejected the concept of a flat Earth as a result of his observations of lunar eclipses. The appropriate figure from Chapter 2 is repeated in Figure 3.3.5.

Sun   Earth   Moon

Figure 3.3.5 Observation leading to the rejection by Aristotle of a flat Earth.

He reasoned that, if Earth was flat, it would cast a straight-edged shadow across the face of the moon during all lunar eclipses. Actual observations revealed no such edge, but rather a shadow with a curved boundary. Therefore, the Earth cannot be flat. (The extended conclusion of a spherical Earth comes by elimination – the only candidate shapes discussed seriously by the ancient philosophers were that of a sphere or plate.) This argument takes on the syllogistic form:

Premise: If Earth is flat, then a straight-edged shadow appears on the moon during a lunar eclipse.

Premise: <u>A straight-edged shadow does not appear on the moon during an eclipse.</u>

Conclusion: ∴ Earth is not flat.

To analyze this using symbolic logic, let p represent, "Earth is flat," and let q represent, "A straight-edged shadow appears on the moon during a lunar eclipse." The corresponding statement form for the syllogism is then

$$[(p \Rightarrow q) \land \sim q] \Rightarrow \sim p$$

which is shown in Figure 3.3.6 to be a tautology. Indirect reasoning is also known by the label *modus tollens*, which we will abbreviate as *MT*.

p	q	$[(p \Rightarrow q) \land \sim q] \Rightarrow \sim p$						
T	T	T T T F	F	**T**	F
T	F	T F F F	T	**T**	F
F	T	F T T F	F	**T**	T
F	F	F T F T	T	**T**	T

1 2 1 3 1 4 1

Figure 3.3.6 This syllogism is a valid argument known as *modus tollens*.

Not all syllogisms are valid arguments as shown in the next example.

Example 3

Construct a truth table to determine the validity of this syllogism.

Premise: If you live in Cuba, you are governed by Communists.
Premise: <u>You are governed by Communists.</u>
Conclusion: $\therefore$ You live in Cuba.

Solution

Let p represent, "You live in Cuba," and q represent, "You are governed by Communists." Then the statement form for this argument is, $[(p \Rightarrow q) \land q] \Rightarrow p$. The truth table is displayed in Figure 3.3.7 and shows that the form is not a tautology. Hence the argument is invalid. ◆

p	q	$[(p \Rightarrow q) \land q] \Rightarrow p$						
T	T	T T T T T	**T**	T
T	F	T F F F F	**T**	T
F	T	F T T T T	**F**	F
F	F	F T F F F	**T**	F

1 2 1 3 1 4 1

Figure 3.3.7 Truth table for an invalid syllogism.

110

Many syllogisms have a categorical nature to them which allow them to be readily analyzed by pictorial diagrams known as **Venn diagrams** in honor of their originator, **John Venn** (1834–1923). Venn was a professor of logic at Cambridge for four decades who made significant contributions to the formal structuring of symbolic logic initiated by George Boole. We illustrate with an example of a *MP* syllogism.

Example 4

Premise:	All flying animals have wings.
Premise:	A robin flies.
Conclusion:	∴ A robin has wings.

Here the statement, "All flying animals have wings," states that the category or set of "flying animals" is contained in the larger set of "winged animals" as a subset. This can be expressed visually by means of a Venn diagram. (To be discussed further in the chapter on probability.) Based on the *first* premise, we draw a small circle to represent the set of all flying animals and place it inside a larger circle which represents the set of all winged animals. (See Figure 3.3.8.) The placement of the specific animal, "robin", is then based on the *second* premise. Since a robin can fly, it is contained in the smaller set which forces it to be a member of the larger set as well. The conclusion that a robin has wings is then consistent with the Venn diagram **which has been drawn based only on the premises**. We therefore conclude this this syllogism is a *valid* argument.

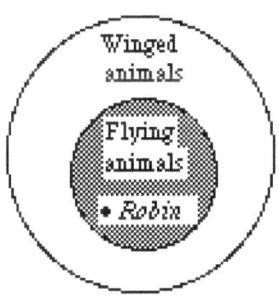

Figure 3.3.8 Venn diagram of an *MP* syllogism.

We can identify this as an *MP* syllogism by first re-wording it as follows:

Premise:	If an animal flies, then it has wings.
Premise:	A robin flies.
Conclusion:	∴ A robin has wings.

We let p represent, "An animal flies," and q represent, "An animal has wings." Note that the use of the word "animal" serves as a variable in this case. Once "robin" has been substituted for "animal", we may write the corresponding statement form as, $[(p \Rightarrow q) \wedge p] \Rightarrow q$, which is the characteristic form of a type *MP* syllogism. ♦

Example 5

Draw a Venn diagram which corresponds to the following *MT* syllogism.

All NFL linemen are at least 6'3" tall.
Patrick is less than 6'3" tall.
∴ Patrick is not an NFL lineman.

Solution

The first premise implies that the set of NFL linemen must be contained in the set of all people who are at least 6'3" tall. (See Figure 3.3.9.) The second premise then requires that Patrick be placed outside the larger set, which in turn places him outside the interior set as well. Only after the diagram has been drawn do we consider the conclusion. Since the conclusion is indeed supported by the Venn diagram, we may conclude that this is a valid argument.

The corresponding statement form, $[(p \Rightarrow q) \wedge {\sim}q]$ is obtained by letting p represent, "A person is an NFL lineman," letting q represent, "A person is at least 6'3" tall," and substituting "Patrick" for "person". This identifies the syllogism type as *MT*.

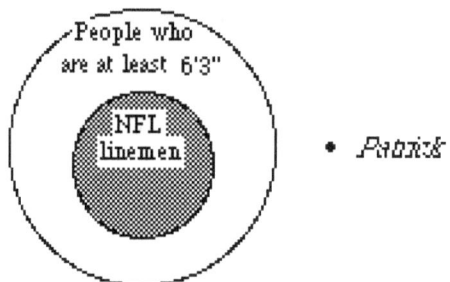

Figure 3.3.9 Venn diagram of an *MT* syllogism.

Example 6

Draw a Venn diagram of this syllogism and determine its validity.

All squares have four right angles.
A rectangle has four right angles.
∴ A rectangle is a square.

Solution

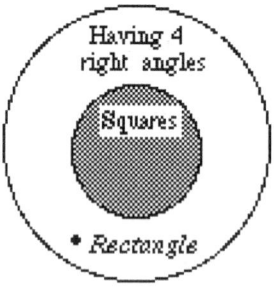

Figure 3.3.10 Venn diagram of an invalid syllogism.

Using only the premises, it is possible in the Venn diagram to place the point associated with the rectangle inside the larger set yet outside the smaller one. Since the conclusion is not

112

supported by a diagram which is consistent with the premises, this syllogism must be declared invalid. The above conclusion is not guaranteed by the premises. If you wished to translate this syllogism to symbolic language, you first observe that it could be re-worded as:

> If a polygon is a square, then it has four right angles.
> A rectangle has four right angles.
> ∴ A rectangle is a square.

Let p represent, "A polygon is a square," and q represent, "A polygon has four right angles." Once you substitute "rectangle" for "polygon", you obtain the form, $[(p \Rightarrow q) \land q] \Rightarrow p$ which has already been demonstrated in Figure 3.3.7 to not be a tautology. Both techniques verify that possession of four right angles does not ensure that a polygonal figure is a square. Either procedure demonstrates that the given argument is invalid. ♦

It is important to be able to distinguish the difference between argument validity and the truth values of the constituent statements. It is similar to our previous observations of true conditional statements having antecedents and consequents which were both false. Consider these examples.

Example 7

All winged animals can fly.	All Senators are over 36 years old.
An ostrich has wings.	Everett is not over 36.
∴ A ostrich can fly.	∴ Everett is not a Senator.

These examples illustrate the irrelevance of the truth of either a premise or the conclusion – taken individually – when determining the *validity* of an argument. It is a fact, for instance, that an ostrich cannot fly. Likewise, any truth about the ages of Senators varies with time. However, given that the premises of each argument are true, the truth of each conclusion is then inescapable according to the rules of logic. The first syllogism is of type *MP* and the second is of type *MT*. ♦

We close with a final type of categorical example conducive to analysis by Venn diagrams.

Example 8

The use of the word "not" in the first premise of each of the following syllogisms implies a separation of sets in the associated Venn diagram. Determine the validity of the argument produced by each of the four variations of the second premise.

I. Texans do not eat carrots.	**II.** Texans do not eat carrots.
Erika is a Texan.	Erika eats carrots.
∴ Erika does not eat carrots.	∴ Erika is not a Texan.
III. Texans do not eat carrots.	**IV.** Texans do not eat carrots.

113

Erika is not a Texan.	Erika does not eat carrots.
∴ Erika eats carrots.	∴ Erika is a Texan.

Solution

In both of the first two syllogisms, the given conclusion follows from the Venn diagrams produced by the premises. Both of these arguments are valid. However, no such thing can be said for the next two cases, because the second premise only directs us to place Erika *outside* of a specific set but not necessarily *inside* the other set. Hence, both of these arguments are invalid.

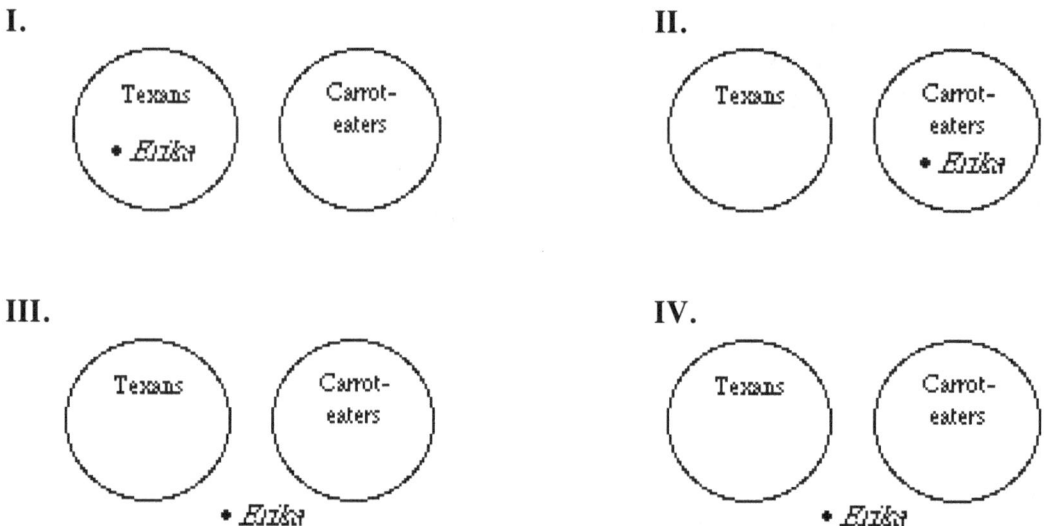

Example 9

One of the most common mistakes of logic made by many people is to conclude that *all* members of one particular set are also members of some second set only because that is the case for one specific member. This mistake is the source of much prejudice throughout human history. So, for instance, just because one person of some faith or occupation or cultural background possesses some characteristic does not imply that all people of that same group share that characteristic. Is the following syllogism valid?

Jorgenson was born in Denmark.
Jorgenson dislikes rock and roll music.
∴ All natives of Denmark dislike rock and roll music.

Solution

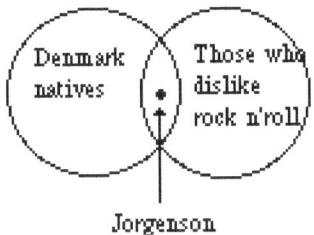

Jorgenson

Clearly this is an invalid argument. The premises only imply that the intersection of these two sets must be non-empty. They do not imply that one set is a subset of the other. So we can draw a diagram that is consistent with the premises but does not support the conclusion. ♦

We close this section by quoting a poignant passage from Leo Tolstoy's novella *The Death of Ivan Ilyich* in which a man is first confronting his inner feelings about the hard reality of his imminent death. We include this here to acknowledge the simple fact that all too often logic is not enough to tackle all of life's issues.

> *In the depth of his heart he knew he was dying, but not only was he unaccustomed to such an idea, he simply could not grasp it, could not grasp it at all. The syllogism he had learned from Kiesewetter's logic – "Caius is a man, men are mortal, therefore Caius is mortal" – had always seemed to him correct as applied to Caius, but by no means to himself. That man Caius represented man in the abstract, and so the reasoning was perfectly sound; but he was not Caius, not an abstract man; he had always been a creature quite, quite distinct from all the others. . . Caius really was mortal, and it was only right that he should die, but for him, Vanya, Ivan Ilyich, with all his thoughts and feelings, it was something else again.*
>
> Leo Tolstoy

Exercise Set 3.3

Determine whether each of the following statement forms is a tautology.

1. $p \vee \sim p$

2. $p \vee q \Leftrightarrow p$

3. $(p \wedge q) \vee (p \vee q)$

4. $(p \wedge q) \Rightarrow p$

5. $[(p \Rightarrow q) \wedge q] \Rightarrow p$

6. $[(p \wedge q) \wedge \sim p] \Rightarrow \sim q$

7. $\sim(p \wedge q) \Rightarrow \sim p \vee \sim q$

8. $\sim(p \vee q) \Rightarrow \sim p \wedge \sim q$

9. $\sim(p \wedge q) \Rightarrow \sim p \wedge \sim q$

10. $\sim(p \vee q) \Rightarrow \sim p \vee \sim q$

11. $[(p \vee q) \wedge \sim p] \Rightarrow q$

12. $[(p \vee q) \wedge p] \Rightarrow \sim q$

13. $[(p \Rightarrow q) \vee (p \Rightarrow r)] \vee (q \Rightarrow r)$

14. $(p \wedge q) \vee (\sim p \wedge \sim q)$

15. $[(p \Rightarrow q) \wedge r] \Rightarrow [p \Rightarrow (q \wedge r)]$

16. $(p \Rightarrow q) \wedge (q \Rightarrow p) \Leftrightarrow (p \Leftrightarrow q)$

Each of the following algebraic expressions has a constant logical value. What is it?

17. $x \geq 3$ or $x < 3$

18. $7x + 2y = 10$ or $7x + 2y \neq 10$

19. $(x + 4)^2 \geq 0$

20. $x > 35$ and $x < 30$

21. $x + 3y > 6$ and $x + 3y \leq 6$

22. $(x > 25$ or $x \leq 25)$ and $|x| < 0$

23. $(x > 1$ and $y > 1)$ and $x + y = 2$

24. $a = 18$ or $(a - 18)^2 > 0$

*25. $\sin^2\theta + \cos^2\theta = 1$ or $\sin\theta = 0.5$

*26. $\sin\theta > 0$ and $\cos\theta = 2$

*Knowledge of trigonometry necessary for these problems.

Create a statement form corresponding to each of the following syllogisms and use it to determine its validity. When appropriate, classify the syllogism as modus ponens *(MP) or* modus tollens *(MT).*

27. If it snows, then Amy builds a fort.
<u>It is snowing.</u>
∴ Amy is building a fort.

28. If Frederich practices his violin, then he gets dessert after supper.
<u>Frederich is not getting dessert.</u>
∴ He did not practice his violin.

29. If you do not watch television, you read books.
You read books.
∴ You do not watch television.

30. If the US supports NATO, then it commits military support to Bosnia.
The US does not support NATO.
∴ The US does not commit military support to Bosnia.

31. Consumption increases if disposable income increases.
Consumption is not increasing.
∴ Disposable income is not increasing.

32. Every time Bert goes fishing, it rains.
It is not raining.
∴ Bert is not fishing.

33. Whenever a dog eats dirt, he gets sick.
Andy's dog does not eat dirt.
∴ Andy's dog does not get sick.

34. Whenever white bugs appear, a tomato spoils.
A tomato is spoiling.
∴ White bugs are appearing.

35. Mold forms whenever the humidity increases.
The humidity is decreasing.
∴ The mold is not forming.

36. New home construction decreases whenever interest rates increase.
New home construction is decreasing.
∴ Interest rates are increasing.

Draw a Venn diagram for each of the following syllogisms and determine whether the syllogism constitutes a valid or invalid argument.

37. All men are mortal.
Caius is a man.
∴ Caius is mortal.

38. All parallelograms have two pairs of equal sides.
A rectangle is a parallelogram.
∴ A rectangle has two pairs of equal sides.

39. All men are mortal.
Caius is mortal.
∴ Caius is a man.

40. All rectangles have four right angles.
A triangle does not have four right angles.
∴ A triangle is not a rectangle.

41. All stars contain hydrogen.
The moon does not contain hydrogen.
∴ The moon is not a star.

42. Every planet orbits the sun.
Halley's comet orbits the sun.
∴ Halley's comet is a planet.

43. Everyone who smokes cigarettes has an above average heart-rate.
Shelly does not smoke cigarettes.
∴ Shelly does not have an above average heart-rate.

44. Humans are all capable of rational thought.
A chimpanzee is not human.
∴ A chimpanzee is not capable of rational thought.

45. Dinosaurs did not have fur.
A woolly mammoth had fur.
∴ A woolly mammoth was not a dinosaur.

46. Accountants never work less than 40 hours/week.
Tracy works 45 hours/week.
∴ Tracy is an accountant.

117

47. No resident of Cleveland is a fan of the Pittsburgh Steelers.
<u>Rick does not live in Cleveland.</u>
∴ Rick is a fan of the Steelers.

48. Nobody over 70 goes skiing.
<u>Georgia is a skier.</u>
∴ Georgia is not over 70.

49. Thomas is from Wisconsin.
<u>Thomas plays the clarinet.</u>
∴ Everyone from Wisconsin plays the clarinet.

50. Omar is from Syria.
<u>Omar is a farmer.</u>
∴ All Syrians are farmers.

51. Ishtak is a lawyer.
<u>Ishtak plays tennis.</u>
∴ All lawyers play tennis.

52. Vito has brown eyes.
<u>Vito is Catholic.</u>
∴ All Catholics have brown eyes.

If possible, deduce a conclusion from the given premises (other than repeating a premise) using valid reasoning.) If this cannot be done, then write, "No conclusion is possible.

53. If you study mathematics more than six hours per week, you will not fail math class. Alexander studies mathematics for 8 hours per week. Therefore, _____

54. If Karolyn plays poker during the week, she borrows money on Friday. Karolyn did not borrow money on Friday. Therefore, _____

55. If the defendant is innocent, then he is not convicted. The defendant is guilty. Therefore, _____

56. All Finns have blue eyes. Leonard has blue eyes. Therefore, _____

57. If the density of a substance exceeds 1.0 gm/cm^3, then the substance sinks. The cork from that wine bottle floats. Therefore, _____

58. All Republican Senators support budget amendment #22. Senator Bureaucrat does not support budget amendment #22. Therefore, _____

59. All Republican Senators support budget amendment #22. Senator Snort is not a Republican. Therefore, _____

60. All piano players are musicians. Constantine plays the guitar. Therefore, _____

61. Everyone who eats spinach is strong. Popeye is not strong. Therefore, _____

62. No one from Luzerne County likes Jazz music. Gilford is not from Luzerne County. Therefore, _____

63. History majors never take more than one mathematics class. Khantana has taken three mathematics classes. Therefore, _____

3.4 Computer Programming Structures

Emotions are illogical.

Mr. Spock, *Star Trek*

When Mr. Spock spoke in the original television series *Star Trek* , his voice usually contained none of the inflections commonly associated with human emotion – a trait shared by the humanoid Data in the follow-up series, *Star Trek: The Next Generation*. This endowed both characters with a personality that most of us would say was, at least in part, rather inhuman. At the same time, the essence of both of these characters centered on their ability to use pure logic in their problem-solving methods. This correspondence of logic and cold detachment stems from most peoples' observations of **computers** – machines designed to perform numerical computations and process information. Why are these machines so good at employing logic?

The answer is connected to electricity. Electricity empowers the logic center of a computer because of the simple fact that, at any given time, an electrical circuit is in one of two states:

Figure 3.4.1 Electricity runs a computer.

on or *off*. Similarly, every storage unit in a computer's memory called a **byte** is typically composed of 8 magnetized **bits** each of which must be either *positive* or *negative*. We say such objects have a **binary** (two-valued) existence and it is this quality that makes orchestrated collections of electrical circuits (like in a computer) so appropriate for implementing logical processes. We have already seen that effective logical analysis is rooted in the fact that statements have a binary quality – they are either true or false. Additionally, each connective is really a binary-valued *function* whose values may be synthesized in a computer by the use of an appropriate mechanism called a **logic gate**.

The main gates used for logic control in a computer are for conjunction (AND) , disjunction (OR), and negation (NOT). The AND and OR gates each have two input terminals and one output terminal, while the NOT gate has just one input terminal and one output terminal. By identifying an abstract statement with a physical electrical conduit, we can associate the value of the statement as "true" with a pulse being sent through the conduit (i.e. *on*) and the value of "false" with the absence of a pulse (i.e. *off*). In order to conform to the truth table for conjunction, the AND gate must be constructed according to the specifications depicted in Figure 3.4.2. Notice how this correlates to the values which $p \wedge q$ acquires as a result of the four combinations of logical values for p and q.

119

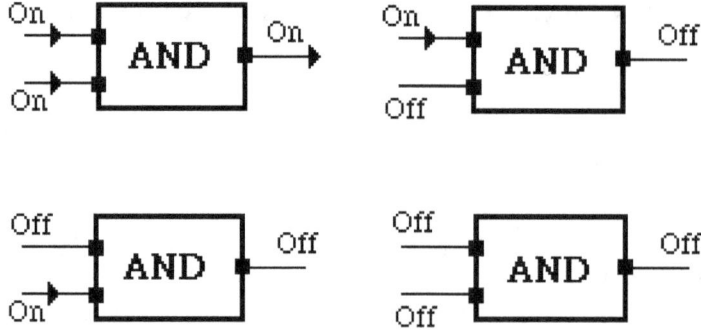

Figure 3.4.2 Current flow through the AND gate.

Similarly, the truth tables for disjunction and negation mandate that the OR and NOT gates be constructed according to the diagrams in Figure 3.4.3.

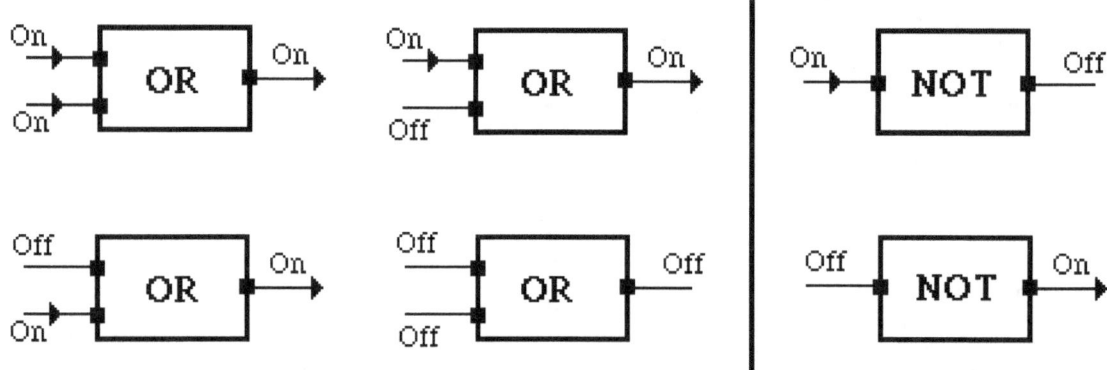

Figure 3.4.3 Current flow through the OR gate and the NOT gate.

A *circuit* is a closed loop of electrical conduit running between a power source and an appliance such as a light bulb. If a switch is inserted in the loop, then the light will be *on* if the switch allows a pulse through and *off* if it does not allow a pulse. For our purposes, we define a **logic circuit** to be a combination of electrical conduits and connective gates that corresponds to a single statement form and acts like one giant switch. If the output terminal of the logic circuit is on, then this corresponds to the form being true. If it is off, then the form is false. The next example illustrates this process.

Example 1

Diagram a logic circuit that corresponds to the statement form, $\sim(p \vee q) \wedge \sim r$, given that p and q both have value T and r has value F. Is the circuit on or off?

Solution

Naturally, the order of evaluation of the connectives is crucial in setting up the circuit in Figure **3.4**.4. We see that this circuit will be off if p, q, and r are initially T, T, and F (on, on, and off respectively). This means that the corresponding statement form, $\sim(p \vee q) \wedge \sim r$, is false.

Clearly, changing the values of the prime statements corresponds to a change in the initial states of the pulses into the circuit which in turn can affect the final state of the circuit. ♦

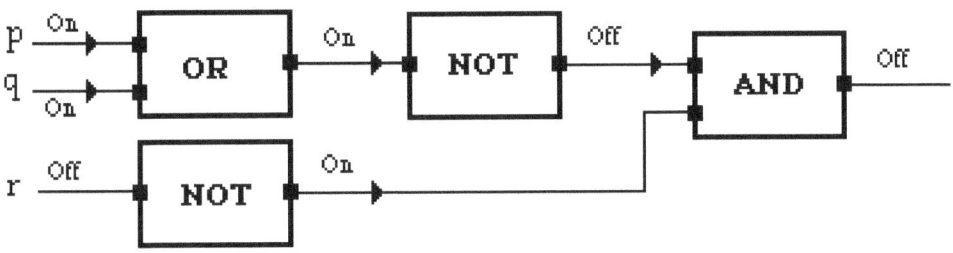

Figure 3.4.4 Logic circuit corresponding to the form, ~(p ∨ q) ∧ ~r.

Exercise Set 3.4

Diagram a logic circuit that corresponds to each of the following statement forms. Given that all the prime statements are true, determine whether the circuit would be on or off.

1. ~*p* ∨ *q*

2. ~(*p* ∨ *q*)

3. *p* ∧ ~(*q* ∧ *r*)

4. (*p* ∧ ~*q*) ∧ ~*r*

5. (~*p* ∨ ~*q*) ∨ *r*

6. (*p* ∧ *q*) ∧ ~(*r* ∧ *s*)

7. *p* ∧ ~*q* ∨ ~(*r* ∧ ~*s*)

8. ~*p* ∧ (*q* ∨ *r*) ∧ ~*s*

9. (~*p* ∨ ~*q* ∧ *r*) ∧ *s*

10. ~(*p* ∨ *q*) ∧ ~(*r* ∨ *s*)

Foxtrot by Bill Amend

Chapter Glossary

algebraic statement   Statement containing a variable.

algorithm   A finite, ordered list of steps for solving a problem.

antecedent   In the conditional statement $p \Rightarrow q$, p is the antecedent.

argument   Collection of a sequence of statements.

biconditional   The connective, "if and only if".

binary   Having exactly two possible values.

bit   The smallest computer memory unit. It is binary-valued.

boolean expression   Computer programming name for an algebraic statement. It is either
true   or false.

byte   Computer memory storage unit usually composed of 8 bits.

cell   Computer memory storage unit composed of one or more bytes.

central processing unit The brain of a computer which performs the arithmetic and logical
                            operations.

compound statement   Statement consisting of two or more prime statements.

computer   Machine which performs numerical computations and processes
                            information.

conclusion   Final statement in an argument.

conditional   The connective, "if . . . then".

conjunction   The connective, "and".

connective   Word used to combine two statements into a compound statement.

consequent   In the conditional statement $p \Rightarrow q$, q is the consequent.

contrapositive   The contrapositive of $p \Rightarrow q$ is $\sim q \Rightarrow \sim p$.

converse   The converse of $p \Rightarrow q$ is $q \Rightarrow p$.

decision structure	Programming method for making a decision.
deduction	Process used to reach a conclusion from a given set of premises in a valid argument.
disjunction	The connective, "or".
equivalent have	Two statement forms are (logically) equivalent if their truth tables identical final columns.
input device	Computer component which transmits information to the memory.
interpretation	A substitution of English words for the symbols in a statement form.
logic circuit	Combination of logic gates which simulates a statement form.
logic gate	Computer mechanism which simulates a logical connective.
logical value	True or false.
looping	Program structure for repeating a set of steps.
machine language	The program instructions in a special language which the computer understands.
modus ponens	Valid reasoning scheme with form $[(p \Rightarrow q) \wedge p] \Rightarrow q$.
modus tollens	Valid reasoning scheme with form $[(p \Rightarrow q) \wedge \sim q] \Rightarrow \sim p$.
negation	Reversal of the logical value of a statement with some form of "not".
output device	Computer component which displays results from the performance of a computer program.
paradox	Sentence or set of conditions leading to a sentence with an undecidable logical value.
pixel	Smallest element of a cathode ray tube.
premise	Any of the statements of an argument except for the conclusion.
prime statement	Statement which does not contain a connective.
program	List of intructions for a computer to translate to machine language and then perform.

pseudocode   Description of an algorithm which the programmer designs before writing the program.

statement   Declarative sentence which has exactly one logical value.

statement form   Any expression of statement symbols and connectives.

symbolic logic   The science of using formal abstractions and procedures to systemize the principles of deduction and valid reasoning.

syllogism   An argument consisting of two premises and a conclusion.

tautology   A statement form (or its interpretation) which is true for every possible combination of logical values for its prime statements.

truth table   A table for listing the logical value of a statement form for every combination of values for the prime statements.

valid   An argument is valid if the truth table corresponding to its statement form is a tautology.

Venn diagram   Pictorial diagram using circles for sets initiated by John Venn used to analyze the validity of arguments.

Chapter Review Test

1. Which of the following sentences are statements?
i) Please mow the front lawn.
ii) Russia is the largest country in the world.
iii) Are you going to Europe this summer?
iv) Jack Nicholson is the greatest actor ever.
v) The US inflation rate was less than 4% in 1992.
vi) Jupiter has more than 12 moons.
vii) This sentence is true.
viii) Mathematics is wonderful!

2. Let p represent, "The high school graduation rate exceeds 90%." Let q represent, "The disciplinary code is strictly enforced." Translate the following statements into symbolic form.

i) The high school graduation rate exceeds 90% and the disciplinary code is strictly enforced.
ii) it is not the case that either the high school graduation rate exceeds 90% or the disciplinary code is strictly enforced.
iii) The high school graduation rate does not exceed 90% whenever the disciplinary code is not strictly enforced.

3. Determine the logical value of the following statements.
i) Abraham Lincoln was the 16th U.S. President or Christmas is on June 25.
ii) Abraham Lincoln was the 16th U.S. President and Christmas is on June 25.
iii) If $45 + 30 = 75$, then whales can fly.
iv) If there are 50 seconds in one minute, then this chapter is more than 1000 pages long.
v) Canada is south of America if and only if the Mississippi River is in California.

4. Translate the following forms into interpretations given these representations.
p: Michael studies 30 hours per week. q: Michael gets good grades.
r: Michael goes to a party on Saturday.

i) $p \lor \sim q$ ii) $p \Rightarrow (q \land r)$ iii) $\sim(p \land r)$

5. Let p represent the algebraic expression, "$x > 0$," and let q represent, "$x + y = 10$." Fill in the following table. Use **T** for true and **F** for false.

x	y	p	q	$L(p)$	$L(q)$	$L(p \land q)$	$L(p \lor q)$
-8	10						
-1	11						
1	7						
5	5						

6. Construct the truth table for this statement form: $(p \land q) \Leftrightarrow \sim(p \lor q)$

7. Determine whether this statement form is a tautology: $(p \land q) \Rightarrow (p \lor q)$

8. Given that p is true, determine whether each statement form has a unique logical value and, if so, find it.

i) $p \lor q$ ii) $p \Leftrightarrow (\sim p \land q)$

9. Use De Morgan's Laws to find the negation of each of these algebraic statements.

  i) $x + y = 17$ and $5z > 21$ ii) $A \neq 100$ or $B \leq 200$

10. Create a statement form that corresponds to the following syllogism and use it to determine its validity.

  If LeBron James scores at least 40 points, then the Miami Heat win.
  <u>The Miami Heat lost.</u>
  ∴ LeBron James scored less than 40 points.

11. Draw a Venn diagram to determine the validity of the following syllogism.

  All humans are sentient creatures.
  <u>An orangutan is not human.</u>
  ∴ An orangutan is not sentient.

12. *If possible*, deduce the correct conclusion using valid reasoning. If you cannot reach a conclusion using valid reasoning, then write, "No conclusion possible."

i) All of the planets orbit the sun in an elliptical path. Mercury is a planet. Therefore, _____

ii) Usain Bolt runs the 100-meter sprint in less than 10 seconds whenever he is healthy. Usain Bolt ran the 100-meter sprint in 9.8 seconds. Therefore, _____

13. Diagram a logic circuit that corresponds to the statement form, $(p \lor \sim q) \land r$. Given that all the prime statements are true, determine whether the circuit would be on or off.

Solutions to Odd-Numbered Exercises

Section 3.1

1. (i), (iv), and (v) are statements.

3.

x	p	$L(p)$
0	$40 \leq 50$	True
3	$46 \leq 50$	True
6	$52 \leq 50$	False
9	$58 \leq 50$	False
12	$64 \leq 50$	False

5.

m	n	p	$L(p)$
0	10	$100 = 100$	True
3	5	$34 = 100$	False
6	6	$72 = 100$	False
6	8	$100 = 100$	True
7	7	$98 = 100$	False
8	6	$100 = 100$	True
9	1	$82 = 100$	False
10	0	$100 = 100$	True

7.   i) I do not ride my bike everyday.

ii) I ride my bike everyday or I weigh less than 200 pounds.

iii) If I ride my bike everyday, then I weigh less than 200 pounds.

iv) If I weigh less than 200 pounds, then I ride my bike everyday.

v) I do not ride my bike everyday and I weigh less than 200 pounds.

vi) It is not the case that I ride my bike everyday and I weigh less than 200 pounds.

vii) I ride my bike everyday if and only if I do not weigh less than 200 pounds.

viii) It is not the case that I ride my bike everyday if and only if I weigh less than 200 pounds.

9.   i) $q \Rightarrow r$ ii) $q \wedge \sim p \wedge \sim r$ iii) $\sim p \wedge \sim q$ iv) $\sim r \Rightarrow \sim q$

v) $\sim p \Rightarrow (q \vee r)$ vi) $p \Leftrightarrow \sim q$ vii) $r \Leftrightarrow (\sim p \wedge q)$ viii) $\sim(p \vee (q \wedge r))$

11.   i) "A polygon is a square," is the antecedent. "A polygon is a rectangle," is the consequent.

ii) "A number is rational," is the antecedent. "A number is real," is the consequent.

iii) "An electric current flows through a wire," is the antecedent . "A magnetic field is created," is the consequent.

iv) "His movies gross $50 million or are seen by 10 million people," is the antecedent. "His movies make a profit," is the consequent.

v) "The topic is gun control or welfare," is the antecedent. "Much debate occurs in Congress," is the consequent.

vi) "It doesn't rain or snow," is the antecedent. "We go fishing," is the consequent.

13. i) $p \vee \sim r$ ii) $(\sim p \wedge q) \vee r$ iii) $p \wedge \sim q \wedge \sim r$

iv) $\sim p \vee \sim (q \wedge r)$ v) $(p \wedge q) \vee (\sim q \wedge r)$

15. No. It cannot be assigned a consistent logical value.
17. Yes.

Section 3.2

1. The statements, "Nick enjoys baseball," and "Nick enjoys spinach," must both be true.
3. False ; True.
5. True ; False.

7.

x	p	q	$L(p)$	$L(q)$	$L(p \wedge q)$	$L(p \vee q)$
500	$500 > 1000$	$500 < 1500$	F	T	F	T
1000	$1000 > 1000$	$1000 < 1500$	F	T	F	T
1250	$1250 > 1000$	$1250 < 1500$	T	T	T	T
1499	$1499 > 1000$	$1499 < 1500$	T	T	T	T
1500	$1500 > 1000$	$1500 < 1500$	T	F	F	T
1600	$1600 > 1000$	$1600 < 1500$	T	F	F	T

9.

x	y	p	q	$L(p)$	$L(q)$	$L(p \wedge q)$	$L(p \vee q)$
62	15	$62 \geq 75$	$77 = 100$	F	F	F	F
70	30	$70 \geq 75$	$100 = 100$	F	T	F	T
75	25	$75 \geq 75$	$100 = 100$	T	T	T	T
80	20	$80 \geq 75$	$100 = 100$	T	T	T	T
90	11	$90 \geq 75$	$101 = 100$	T	F	F	T

11. T   **13.** F   **15.** T

17.

p	q	~p ⇒ q
T	T	**T**
T	F	**T**
F	T	**T**
F	F	**F**

19.

p	q	~(p ∧ q)
T	T	**F**
T	F	**T**
F	T	**T**
F	F	**T**

21.

p	p ∧ ~ p
T	**F**
F	**F**

23.

p	q	(p ⇒ q) ∧ (q ⇒ p)
T	T	**T**
T	F	**F**
F	T	**F**
F	F	**T**

25.

p	q	r	(p ∧ q) ∨ r
T	T	T	**T**
T	T	F	**T**
T	F	T	**T**
T	F	F	**F**
F	T	T	**T**
F	T	F	**F**
F	F	T	**T**
F	F	F	**F**

27.

p	q	r	p ⇒ (q ∨ ~r)
T	T	T	**T**
T	T	F	**T**
T	F	T	**F**
T	F	F	**T**
F	T	T	**T**
F	T	F	**T**
F	F	T	**T**
F	F	F	**T**

29. T **31.** Either logical value is possible. **33.** F

35. T **37.** F

39.

p	q	p ⇔ q	(p ⇒ q) ∧ (q ⇒ p)
T	T	T **T** T	T T T **T** T T T
T	F	T **F** F	T F F **F** F T T
F	T	F **F** T	F T T **F** T F F
F	F	F **T** F	F T F **T** F T F

41.

p	q	p $\Rightarrow$ q			~p	$\lor$	q
T	T	T	**T**	T	F	**T**	T
T	F	T	**F**	F	F	**F**	F
F	T	F	**T**	T	T	**T**	T
F	F	F	**T**	F	T	**T**	F

43. Either Senator Farnsworth did not vote for the tax cut or he did not vote for the new spending bill.

45. Erika did not receive a B in Math and she did not receive a C in English.

47. $x \le 2$ or $y \ne 35$

49. $x + y \ne 4$ and $x \ge 13.5$

51. $y \ge 85$ and $y \le 90$

53. $(x \ge 10$ or $y \ge 10)$ and $2x + 3y \ne 50$

55. Converse: If the temperature of the earth's core is between 4000 and 5000° C., then it is made principally of iron.
Contrapositive: If the temperature of the earth's core is not between 4000 and 5000° C., then it is not made principally of iron. ; The contrapositive.

57. Converse: If $f(2) = 19$, then $f(x) = 8x + 3$.
Contrapositive: If $f(2) \ne 19$, then $f(x) \ne 8x + 3$. ; The contrapositive.

*59. $p \Rightarrow q \equiv \sim(p \land \sim q)$

*61. $p \land q \equiv$ **NOR**(**NOR**(p, q), **NOR**(p, q))

Section 3.3

1. Yes	**3.** No	**5.** No	**7.** Yes
9. No	**11.** Yes	**13.** Yes	**15.** Yes
17. True	**19.** True	**21.** False	**23.** False
*25. True	**27.** Valid. *MP*	**29.** Invalid.	**31.** Valid. *MT*
33. Invalid.	**35.** Invalid.		

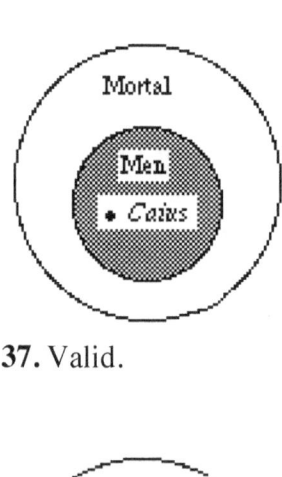

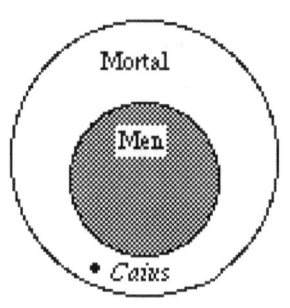

37. Valid.

39. Invalid.

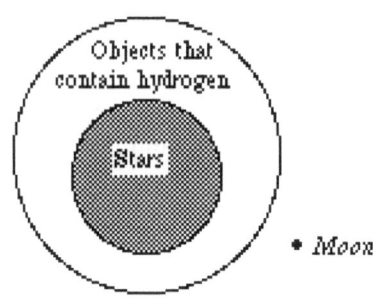

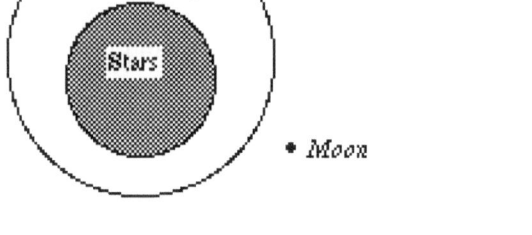

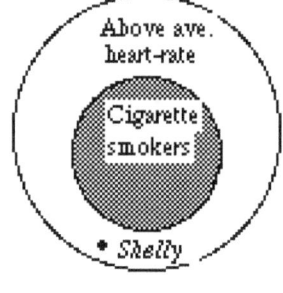

41. Valid.

43. Invalid.

45. Valid.

47. Invalid.

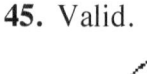

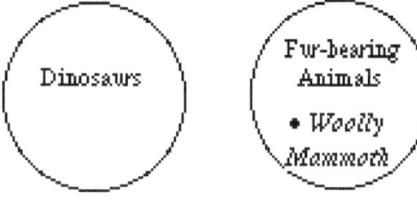

49. Invalid.

51. Invalid.

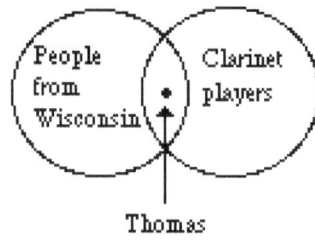

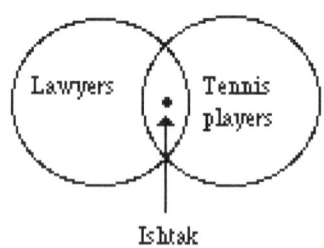

53. Alexander will pass his math class.
55. No conclusion is possible.
57. The cork has a density less than or equal to 1.0 gm/cm^3.
59. No conclusion is possible.
61. Popeye does not eat spinach.
63. Khartana is not a history major.

Section 3.4

1.

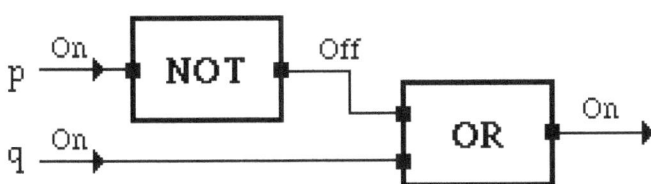

3.

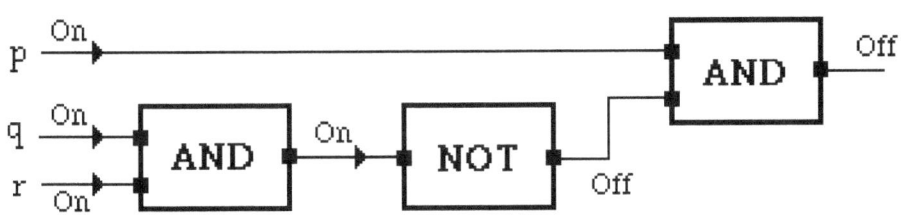

5.

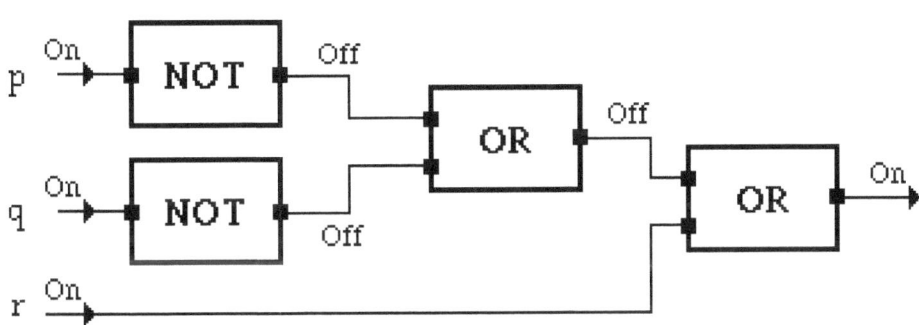

7.

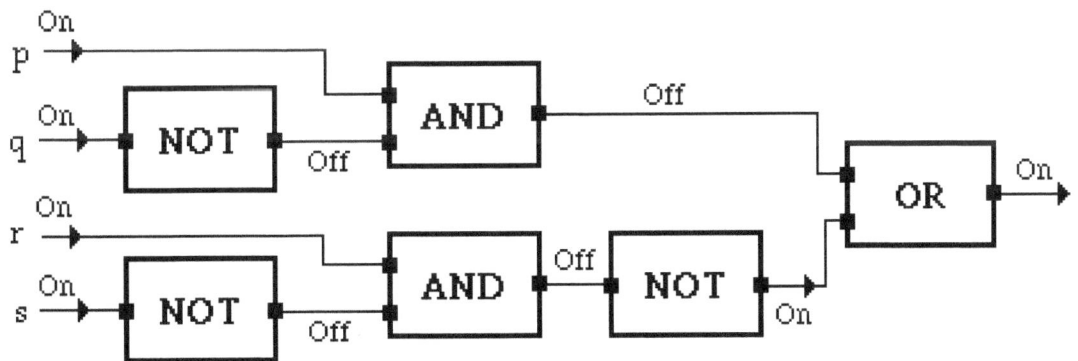

133

9.

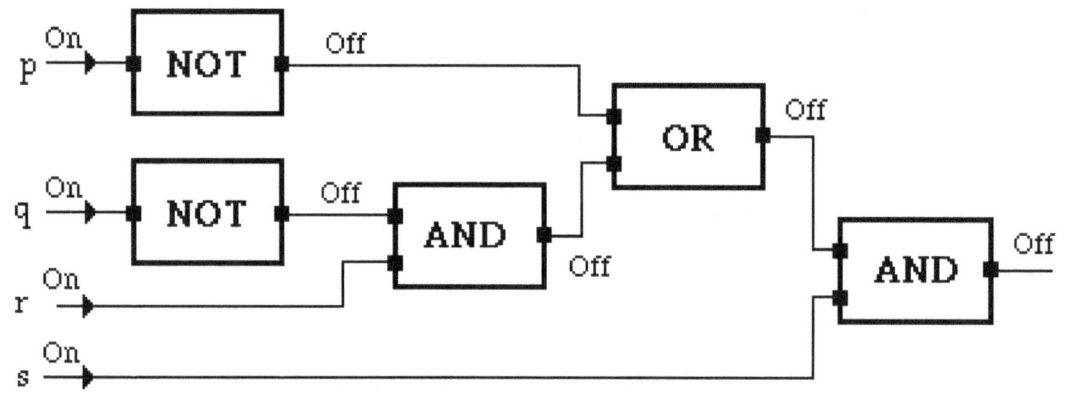

Chapter Review Test

1. (ii), (v), (vi), and (vii) are statements.

2. i) $p \wedge q$ ii) $\sim(p \vee q)$ iii) $\sim q \Rightarrow \sim p$

3. i) T ; ii) F ; iii) F ; iv) T ; v) T

4.   i) Michael studies 30 hours per week or he does not get good grades.

ii) If Michael studies 30 hours per week, then he gets good grades and he goes to a party on Saturday.

iii) It is not the case that Michael studies 30 hours per week and he goes to a party on Saturday.

5.

x	y	p	q	$L(p)$	$L(q)$	$L(p \wedge q)$	$L(p \vee q)$
-8	10	$-8 > 0$	$2 = 10$	F	F	F	F
-1	11	$-1 > 0$	$10 = 10$	F	T	F	T
1	7	$1 > 0$	$8 = 10$	T	F	F	T
5	5	$5 > 0$	$10 = 10$	T	T	T	T
25	-15	$25 > 0$	$10 = 10$	T	T	T	T

6.

p	q	p ∧ q ⇔ ~(p ∨ q)
T	T	T T T **F** F T T T
T	F	T F F **T** F T T F
F	T	F F T **T** F F T T
F	F	F F F **F** T F F F

7. Yes, it is a tautology.   **8.** i) T ii) F

9. i) $x + y \neq 17$ or $5z \leq 21$ ii) $A = 100$ and $B > 200$

10. $[(p \Rightarrow q) \wedge \sim q] \Rightarrow \sim p$. Valid *MT* syllogism.

11. Invalid.

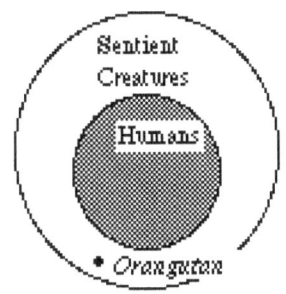

12. (i) Mercury orbits the sun in an elliptical path.   (ii) No conclusion is possible.

13.

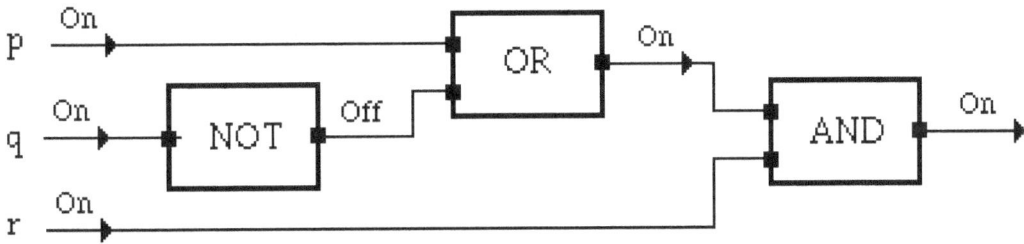

References

1. *Men of Mathematics* by E.T. Bell, Paw Prints, 2008.

2. *The World of Mathematics* by James R. Newman, Dover Publications, 2003

3. *Introduction to Mathematical Logic* by Elliott Mendelson, Chapman & Hall, 5th edition, 2009.

4. *Math and Logic Games* by Franco Agostini, Harper-Collins, 1986.

5. *The Essence of Logic* by John J. Kelly, Prentice Hall, 1996.

6. "A Visual Approach to Deductive Reasoning" by Frances Van Dyke, *Mathematics Teacher*, NCTM, Volume 88, Number 6, September 1995.

7. *The Changing Earth* by J. Monroe and R. Wicander, Brooks-Cole, 4th edition, 2005.

8. *Sets, Logic, and Axiomatic Theories* by Robert R. Stoll, W.H. Freeman, San Francisco, 1979.

9. *New Applications of Mathematics*, Edited by Christine Bondi, Penguin Books, London, 1991.

10. *Oh! Pascal!* by Doug Cooper, Third Edition, W.W. Norton & Company, New York and London, 1993.

11. *Computability, Computable Functions, Logic, and the Foundations of Mathematics* by R.L. Epstein/W.A. Carnielli, Wadsworth & Brooks/Cole, Pacific Grove, CA, 1989.

12. *Introduction to Mathematical Philosophy* by Bertrand Russell, London: George Allen & Unwin, Ltd., New York: The Macmillan Co., 1924.

Chapter 7
Probability

They sit with papers before them scrawled over in pencil, note the strokes, reckon, deduce the chances, calculate, finally stake and – lose exactly as we simple mortals who play without calculations. On the other hand, I drew one conclusion which I believe to be correct: that is, though there is no system, there really is a sort of order in the sequence of casual chances – and that, of course, is very strange.

The Gambler by Fyodor Dostoyevsky

The "order in the sequence of casual chances", as spoken by the protagonist of Dostoyevsky's novel, has been noticed in games of chance and everyday life for centuries. Of course, gamblers are not the only ones interested in being able to effectively make predictions about the future. The intent of this chapter is to provide a brief exposure to how chance and probability are systematically studied and we begin by introducing the basic concept of a *sample space* of outcomes that can result from some well-defined action or experiment.

7.1 Sample Spaces and Events

Suppose you decide to buy 10 shares of your favorite stock tomorrow. Is the price determined? Clearly the answer is no. You could look up today's opening price in the newspaper, but it certainly could change before you buy the stock tomorrow. It could even change while you are on the telephone placing your order. Similarly, do you know what the high temperature will be tomorrow or even whether it will rain? Your answer may depend on where you live and the time of year, but nevertheless no one can be positive about the future state of the weather at any given time. What about your future career, the interest rate at the end of the year, your grade in your next mathematics course, or the roll of two dice? Although you may be able to make a good guess at what might happen, you cannot state with certainty the outcome in any of these cases. They cannot be predicted absolutely in the manner of *deterministic phenomena* such as the effects of gravity or the result of mixing two specific chemicals in the laboratory. Throughout this chapter we will be studying random **experiments**. These are actions or processes for which the outcome cannot be predicted with certainty. Probability is the study of patterns found in repeated occurrences of such experiments. Some examples are listed below.

- Record the price of your favorite stock at the end of the next Tuesday.
- Record the high temperature for the day in your town tomorrow.
- Flip a coin and record the outcome.
- Pick a card from a standard (52 card) deck and record the suit.
- Select five numbered balls from a barrel and record the values.
- Roll two six-sided dice and record the sum of the number of pips showing.
- Pick an integer between 1 and 10, inclusive.
- Do a Google search for a particular topic and write down the number of results.

The above are all examples of random experiments, because in each case the outcomes cannot be predicted with certainty. Notice that you can list the possible outcomes (however unlikely) that may occur. For the coin toss we know the coin must come up heads (H) or tails (T), so we could say the set $\{H, T\}$ lists all possible outcomes from the experiment. A set that lists all the distinct possible outcomes from a random experiment is called a ***sample space***, usually denoted by S. ***Sets***, collections of objects, are a useful tool in the study of probability and we will study them in greater detail in the next section. For now we just examine some sample spaces that correspond to a few different experiments.

Example 1
• Roll two six-sided dice and record the sum of the number of pips showing. Then the sample space can be given by $S = \{2, 3, 4, 5, 6, 7, 8, 9, 10, 11, 12\}$.
• Pick a card from a standard deck and record the suit. Then the sample space can be given by $S = \{$Heart, Spade, Club, Diamond$\}$.
• Pick a card from a standard deck and record the value (denomination) on the face. Then the sample space can be given by $S = \{2, 3, 4, 5, 6, 7, 8, 9, 10,$ Jack, Queen, King, Ace$\}$. ◆

The above examples illustrate an important point you must consider in designing a sample space. The experiment does not only involve an action (drawing a card), but also recording an outcome. In the second experiment we were interested in recording the suit after selecting a card and so we listed suits in our sample space. In the next instance, however, we were interested in recording the denomination on the face and so our sample space consisted of a list of denominations. The next example also illustrates this distinction.

Example 2
   • Toss a fair coin twice and record the outcomes. The sample space can be given by $S = \{HH, HT, TH, TT\}$, where H represents heads and T represents tails.
   • Toss a fair coin twice and record the number of heads. The sample space can be given by $S = \{0$ heads, 1 head, 2 heads$\}$. ◆

Different methods exist for listing the members of the samples space as seen in Example 3.

Example 3
   • Pick an integer between 1 and 10, inclusive. The sample space can be given by $S = \{1, 2, 3, 4, 5, 6, 7, 8, 9, 10\}$, which is the list of integers between 1 and 10, including both 1 and 10.
   • Pick a four-digit number. Though large, we can list this sample space by establishing a pattern. The sample space is $S = \{1000, 1001, 1002, ..., 9998, 9999\}$. Note that we did not list every outcome from this experiment, but we have developed enough of a pattern to fully describe the set.

• Toss a coin and record the number of tosses required until a head is observed. The sample space can be listed as $S = \{1, 2, 3, 4, 5, 6, ...\}$. Notice that this is an infinite set, which is indicated by three dots trailing the list that defines the pattern. ◆

Using a listed set to describe outcomes is not always the most convenient way to write a sample space. Suppose you have two dice: one red and one green. You roll both dice and record the outcome. In this case the most convenient form for a sample space might be a chart like the one in Figure 7.1.1.

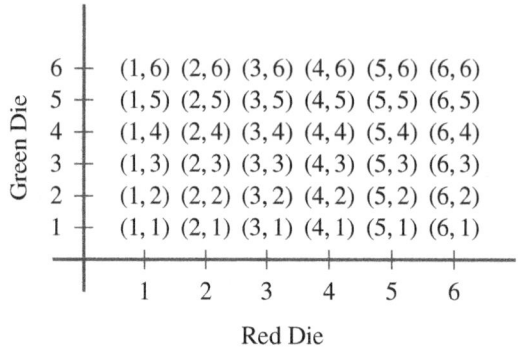

Figure 7.1.1 Two dice chart.

We now turn our attention to measuring a specific set of outcomes from an experiment. Suppose we toss a coin three times and record the outcome. The sample space could be written as $S = \{HHH, HHT, HTH, HTT, THH, THT, TTH, TTT\}$. If we want to know the chances of obtaining two heads out of three tosses, we are now interested in a particular *event*. Since an event is the occurrence of some combination of possible outcomes (including *no* outcomes), it is formally defined as any subset of a sample space. For our current space, suppose we describe event A like this: obtaining *exactly* two heads. According to our definition, we would list $A = \{HHT, HTH, THH\}$. Other events of interest might be obtaining *at least* two heads, which we list as $B = \{HHH, HHT, HTH, THH\}$; obtaining at least one tail, which we list as $C = \{HHT, HTH, HTT, THH, THT, TTH, TTT\}$; or obtaining a head on the first toss and a tail on the second toss, which we list as $D = \{HTH, HTT\}$. We return to our previous examples and describe some typical events.

Example 4

• $S = \{2, 3, 4, 5, 6, 7, 8, 9, 10, 11, 12\}$ for the experiment of rolling two dice and adding the number of pips. Let A be the event that the sum is between 7 and 9, inclusive. We can list $A = \{7, 8, 9\}$. Let B be the event that the sum is odd. We can list $B = \{3, 5, 7, 9, 11\}$.

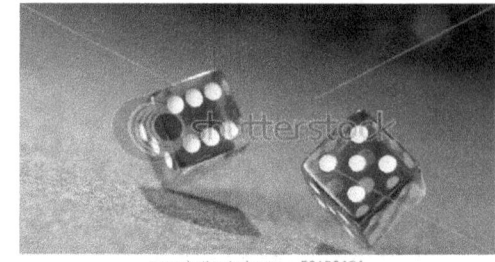

• $S = \{$Heart, Spade, Club, Diamond$\}$ for the experiment of drawing a card from a deck and recording the suit. Let R be the event that the card is red. We can list $R = \{$Heart, Diamond$\}$.

138

• $S = \{2, 3, 4, 5, 6, 7, 8, 9, 10, \text{Jack}, \text{Queen}, \text{King}, \text{Ace}\}$ for the experiment of drawing a card from a deck and recording the denomination. Let F be the event that the card is a face card. We can list $F = \{\text{Jack}, \text{Queen}, \text{King}\}$. ◆

Example 5

• $S = \{HH, HT, TH, TT\}$ for the experiment of tossing two coins. Let E be the event that more than two heads occur. Then $E = \emptyset$ (the empty set), since none of the outcomes in S contain more than two heads.
• $S = \{0 \text{ heads}, 1 \text{ head}, 2 \text{ heads}\}$ for the experiment of tossing two coins and recording the number of heads. Let T be the event that we obtain exactly 1 tail. Since we must also have 1 head, then $T = \{1 \text{ head}\}$. ◆

Example 6

• $S = \{1, 2, 3, 4, 5, 6, 7, 8, 9, 10\}$ for the experiment of picking an integer between 1 and 10 inclusive. Let P be the event that the integer is a prime number. We can list $P = \{2, 3, 5, 7\}$. Recall that a number is prime if it is larger than 1 and is divisible only by itself and 1.
• $S = \{1000, 1001, 1002, ..., 9998, 9999\}$ for the experiment of picking a four-digit number. Let F be the event that the first digit is 5. We can list $F = \{5000, 5001, 5002, ... , 5998, 5999\}$. Note that we do not list every outcome in this event, but the pattern is clearly established.
• $S = \{1, 2, 3, 4, 5, 6, ...\}$ for the experiment of tossing a coin and recording the number of tosses required until a head is observed. Let Q be the event that fewer than four tosses are required. We can list $Q = \{1, 2, 3\}$. ◆

Probability is concerned with measuring the likelihood that an event happens. As such, most probability problems begin with the establishment of a sample space and an event to measure. It is interesting to note a relationship that events have with sample spaces. In Example 4 we listed the event A, that corresponded to the sum of pips on two dice, as $A = \{7, 8, 9\}$. This was based on the sample space $S = \{2, 3, 4, 5, 6, 7, 8, 9, 10, 11, 12\}$. Yet we also established a separate sample space S' using a chart (Figure 7.1.1) that listed the outcomes of rolling two dice. Could S' be used to help define the event A? Using dotted lines, we can note that diagonals correspond to the sums listed in S.

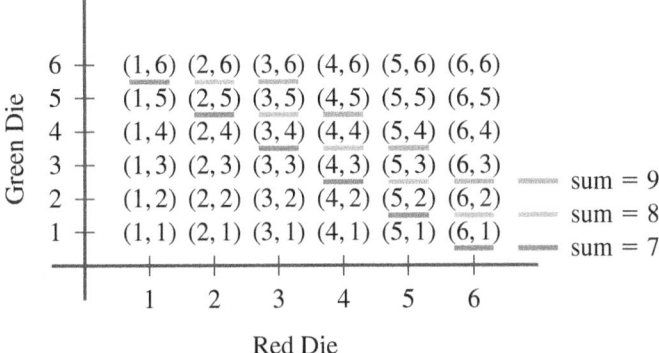

Figure 7.1.2 Two dice chart.

139

So the event $A' = \{(6,1), (5,2), (4,3), (3,4), (2,5), (1,6), (6,2), (5,3), (4,4), (3,5), (2,6),$ $(6,3), (5,4), (4,5), (3,6)\}$ in the sample space S' corresponds to the possible dice combinations resulting in a sum of 7, 8, or 9. Which of these spaces is optimal for use in the area of probability? Typically the sample space best suited to clarify the situation when assigning probabilities is the one in which every outcome is equally likely to occur. This is the case here for S' and moreover S' contains sufficient information to describe a greater number of events than S. For example, we can list the event of obtaining "doubles" as $B = \{(1,1), (2,2), (3,3), (4,4),$ $(5,5), (6,6)\}$ while S does not contain such information. It would not be an appropriate sample space if obtaining "doubles" was of interest. This relationship between events and sample spaces will be revisited in Section 7.3, when we study the probability of an event occurring.

Exercise Set 7.1

1. Two nine-sided dice are rolled (numbered 1 through 9) and the sum of the pips is recorded.

   i) List the sample space *S*.
   ii) List the event *E* that the sum is even.
   iii) List the event *A* that the sum is at least 10.
   iv) List the event *L* that the sum is less than 6.
   v) List the event *P* that the sum is a prime number.

2. One red die and one green die (six-sided) are rolled and their outcomes are recorded.

   i) List the sample space *S*.
   ii) Let *F* be the event that both dice show an odd number of pips. List the event *F*. Be sure your description is consistent with your sample space from (i).
   iii) Let *B* be the event that the sum of pips is more than 8. List the event *B*.
   iv) Let *N* be the event that the number of pips showing on the red die is larger than the green die. List the event *N*.
   v) Let *T* be the event that the number of pips showing on the green die is exactly 2 more than the red die. List the event *T*.
   vi) Let *A* be the event that the number of pips showing on the green die is at least 2 more than the red die. List the event *A*.

3. Repeat Exercise 2, assuming the red die is four-sided and the green die is six-sided.

4. A student is asked to list a sample space for tossing two coins and lists *S* = {0 heads, 1 head, 0 tails, 1 tail}.

   i) Explain why this is not a valid sample space for the experiment.
   ii) List a correct sample space.

5. Give the sample space for each of the following experiments.

   i) A whole number is picked at random between 0 and 100, inclusive.
   ii) A five-digit number is picked at random.
   iii) A coin is tossed three times, and the number of tails is observed.
   iv) A coin is tossed three times, and the sequence of heads and tails is observed.

6. A liberal arts mathematics course contains freshmen, sophomores, juniors, and seniors. A student is picked at random and his/her class and sex are recorded.

   i) List the sample space *S*. (*Hint*: Use ordered pairs.)
   ii) Let *F* be the event that the student is a female. List the event *F*.
   iii) Let *U* be the event that the student is a junior or senior. List the event *U*.
   iv) Let *M* be the event that the student is a freshman or a male. List the event *M*.

7. Three fair coins are tossed and the coin faces are recorded.
    i) List the sample space.
    ii) List the event *A* that at least one head occurs.
    iii) List the event *B* that exactly one tail occurs.
    iv) List the event *C* that exactly two tails occurs.
    v) List the event *D* that exactly no heads occur.

8. A fair coin is tossed four times and the sequence of heads and tails is recorded.

i) List the sample space *S*.
ii) Let *T* be the event that more than two heads are recorded. List the event *T*. Be sure your description is consistent with your sample space listed in (i).
iii) Let *N* be the event that either no heads or no tails are recorded. List the event *N*.
iv) Let *F* be the event that the first tail that appears must be directly preceded by a head. List the event *F*.
v) Let *E* be the event that exactly 3 coins of the same face (e.g., 3 heads or 3 tails) appear. List the event *E*.
vi) Let *A* be the event that no two consecutive tosses result in the same facing. List the event *A*.

9. The American roulette wheel shown below that is usually used in a casino has 38 slots numbered 1, 2, 3, ..., 36, (colored either red or black) as well as the "house slots" 0 and 00 (colored green). A ball is rolled around the wheel and randomly drops into one of the slots.

i) List the sample space *S*.
ii) List the event *R* that the ball drops into a red slot.
iii) List the event *B* that the ball drops into a black slot.
iv) Let *M* be the event that the ball drops into a red slot numbered between 19 and 34.
v) Let *G* be the event that the ball drops into a "house" slot.
vi) Let *E* be the event that the ball drops into an even slot or into a green slot.

10. In the game of blackjack, your objective is to have more points than the dealer without exceeding 21. We can use ordered pairs to represent outcomes. For example, $(19, 17)$ could represent the result that you obtained 19 while the dealer obtained 17 (so you win!). Similarly, $(23, 19)$ would represent the result that you obtained 23 while the dealer obtained 19 (you lose, since you exceeded 21). Because of certain "house rules" for dealers, we'll assume the dealer's points can range between 17 and 26. Due to accepted practices for players, we'll assume your points can range between 12 and 27.

i) Using ordered pairs as described above and set-builder notation, list the sample space S.

ii) You "bust" if your points exceed 21 (regardless of the dealer points). List the event "you bust."

iii) You win if your points exceed the dealer's points (and you don't bust), *or* if the dealer busts while your points remain 21 or lower. List the event "you win."

iv) You "push" if your points are identical to the dealer's, without either of your hands busting. List the event "you push." Blackjack players note: For simplicity, we are representing a "blackjack" result by 21 without any special considerations.

v) You lose if you bust, *or* if your points are less than the dealer's and the dealer doesn't bust. List the event "you lose."

7.2 Introduction to Sets

As Section 7.1 illustrated, the use of sets is essential in studying probability. Recall that the sample space for an experiment is the *set* of all possible outcomes, and that an event is any *subset* of the sample space. In order for us to combine events, therefore, we must first be able to combine sets. Implicit in our discussion is an assumption of the existence of an all-encompassing set for a particular situation. A **universal set** is the collection of all elements for a situation; thus the sample space serves as a universal set in an experiment, whose elements are the possible outcomes of the experiment. In the study of probability, we define a **set** as a collection of elements within a given universal set. For example, let U = {all regularly enrolled college students} be the universal set, and Y = {students enrolled at Yale University}. Then Y is a set, since it is a collection of elements (students) within U. It is important that a universal set be specified for any problem you encounter, just as a sample space is essential for any experiment. A universal set is situation-specific, that is, it may change from problem to problem.

Let U be a universal set and let A and B be sets. We say A is a **subset** of B, denoted $A \subseteq B$, if every element of A is also an element of B. Using the symbol $\in$ for "is an element of", this can be stated: $x \in A$ implies that $x \in B$ if and only if $A \subseteq B$. The set containing no elements is called the **empty set** and is represented by the symbol $\varnothing$. It must necessarily be the case that $\varnothing$ and the universal set U are always subsets of U.

Building on our previous example, consider the sets:

U = {all regularly enrolled college students}
Y = {students enrolled at Yale University}
E = {students majoring in English at Yale}

Is $Y \subseteq E$, $E \subseteq Y$, neither, or both? If $Y \subseteq E$, then every student enrolled at Yale would have to be majoring in English. But this is not true, since there are students enrolled at Yale who have majors other than English. So Y is not a subset of E, which we can denote $Y \nsubseteq E$. Our second question was whether $E \subseteq Y$. If $E \subseteq Y$, then every student majoring in English at Yale must be enrolled at Yale. Since this is true, we can say E is a subset of Y, denoted $E \subseteq Y$. The reasoning used above to show $Y \nsubseteq E$ is important enough to emphasize the following rule:

To show $A \nsubseteq B$, it suffices to find just one element of A that is not an element of B.

Example 1

Let U = {major league baseball players}, L = {left-handed hitters}, and R = {right-handed hitters}. If $L \subseteq R$, then every left-handed hitter would have to be a right-handed hitter also. But this is not true, as there are left-handed hitters who never bat right-handed. So $L \nsubseteq R$. Similar reasoning shows that $R \nsubseteq L$. ◆

Example 2

Let U = {dice combinations from two six-sided die rolls}, D = {doubles are obtained}, E = {sum of dice is even}, and F = {both dice are odd}. Note that $D \nsubseteq F$, since there is a double (for example, $(4, 4)$) in which both outcomes are not odd. Also $F \nsubseteq D$, since there are elements of F (for example, $(1, 5)$) which are not doubles. However $D \subseteq E$, since adding the rolls of doubles produces an even sum. You should check that $E \nsubseteq D$, $F \subseteq E$, and $E \nsubseteq F$. ◆

We now turn our attention to set operations. These are operations that create new sets out of existing sets. The **complement** of set A, denoted $\overline{A}$, is the set of all objects in the universal set which are not in A. Equivalently, we can write, $\overline{A} = \{x \mid x \in U \text{ and } x \notin A\}$. (Note the use of the symbol $\notin$ to represent "is not an element of.") From Example 2, the following complements should be evident:

$$\overline{D} = \{\text{doubles are not obtained}\} = \{\text{the two dice are not equal}\}$$
$$\overline{E} = \{\text{sum of dice is not even}\} = \{\text{sum of dice is odd}\}$$
$$\overline{A} = \{\text{it is not the case that both dice are odd}\} = \{\text{at least one die is even}\}$$

It should also be evident that the complement of a complement is just the original set. Using complement notation this says that $\overline{\overline{A}} = A$, for any set A. Note that it must always be true that $\overline{U} = \varnothing$ and $\overline{\varnothing} = U$.

Two additional set operations with which you may be familiar are union and intersection. The *union* of sets A and B, denoted $A \cup B$, is the set of objects belonging to A *or* B (or both). Equivalently, $A \cup B = \{x \mid x \in A \text{ or } x \in B\}$. The *intersection* of sets A and B, denoted $A \cap B$, is the set of objects belonging to *both A and B*, or symbolically, $A \cap B = \{x \mid x \in A \text{ and } x \in B\}$.

From Example 1, the union of L and R is the entire universal set, since all hitters must bat from one or the other side of the plate. So $L \cup R = U$, the universal set. The intersection of L and R is the set of hitters who bat from both left and right sides (called switch hitters). So $L \cap R = \{\text{switch hitters}\}$. Note that, since $A \cup B$ contains elements in either A or B (or both), then it is always true that $A \cap B \subseteq A \cup B$.

Example 3

Let $U = \{1, 2, 3, 4, 5, 6, 7, 8, 9, 10\}, P = \{2, 3, 5, 7\}, D = \{1, 3, 5, 7, 9\}, E = \{2, 4, 6, 8, 10\}$. Find: $\overline{P}, \overline{D}, P \cap D, \overline{P} \cap D, P \cup D, P \cap E, \overline{P} \cap E, P \cup E, D \cap E, D \cup E$.

Solution

$$\overline{P} = \{1, 4, 6, 8, 9, 10\} \qquad\qquad \overline{D} = \{2, 4, 6, 8, 10\} = E$$
$$P \cap D = \{3, 5, 7\} \qquad\qquad \overline{P} \cap D = \{1, 9\}$$
$$P \cup D = \{1, 2, 3, 5, 7, 9\} \qquad\qquad P \cap E = \{2\}$$
$$\overline{P} \cap E = \{4, 6, 8, 10\} \qquad\qquad P \cup E = \{2, 3, 4, 5, 6, 7, 8, 10\}$$
$$D \cap E = \varnothing \text{ (the empty set)} \qquad\qquad D \cup E = \{1, 2, 3, 4, 5, 6, 7, 8, 9, 10\} = U$$

Note that we only list elements once when performing unions, even if the element appears in both sets. For example, in $P \cup D$ the element 5 is only listed once even though $5 \in P$ and $5 \in D$. ◆

The previous example brings to light several observations. For instance, an intersection can be empty. It is not necessary for two sets to have a common intersection point. In general, two sets that have an empty intersection are called *disjoint*. If these sets represent events in an experiment, then they are called *mutually exclusive events* and such events cannot occur at the same time. In two coin tosses, the events $A = \{\text{obtain two heads}\}$ and $B = \{\text{obtain two tails}\}$ are mutually exclusive events, since $A \cap B = \varnothing$. If instead we let $A = \{\text{obtain at least one head}\}$ and

$B = \{$obtain at least one tail$\}$, note that $A \cap B = \{HT, TH\}$, and so now A and B are not mutually exclusive events.

We also note in Example 3 that although D and E have no elements in common $(D \cap E = \varnothing)$, together they comprise the universal set $(D \cup E = U)$. Here the result is not surprising, since $U = \{$whole numbers from 1 through 10$\}$, $D = \{$odd numbers from 1 through 10$\}$, and $E = \{$even numbers from 1 through 10$\}$. We call D and E a *partition* of U. In general, a pair of subsets A and B of a set E form a **partition** of E if $A \cap B = \varnothing$ and $A \cup B = E$. Notice that any set A and its complement must form a partition of the universal set, since it is always true that

$$A \cap \overline{A} = \varnothing \quad \text{and}$$
$$A \cup \overline{A} = U.$$

We will utilize this idea later in the section.

Example 4

Let U be a standard (52 card) deck of cards. Each of the following pairs of sets form a partition of U.

i) $R = \{$red cards$\}$ and $B = \{$black cards$\}$. Note that $\overline{R} = B$ and likewise $\overline{B} = R$.

ii) $N = \{2, 3, 4, 5, 6, 7, 8, 9, 10\}$ and $P = \{$Jack, Queen, King, Ace$\}$

iii) $E = \{2, 4, 6, 8, 10, \text{Queen, Ace}\}$ and $F = \{3, 5, 7, 9, \text{Jack, King}\}$

Using diagrams is also useful to describe set operations. In particular, pictures known as **Venn diagrams** are particularly useful to illustrate the ideas of complement, union, and intersection. They are named in honor of John Venn who was born in Hull, England in 1834 and attended Gonville and Caius College Cambridge on a mathematics scholarship. Upon graduation he become a Fellow of the College and two years later an ordained priest as well. By 1862 he was a lecturer at Cambridge in moral science, philosophy, and probability and began to find himself drifting from the "orthodox clerical outlook". In the same year in which he was elected to the Royal Society, he left the priesthood, but remained "throughout his life a man of sincere religious conviction".

Figure 7.2.1 John Venn (1834-1923)

In a typical Venn diagram, a rectangle is used to represent the universal set, circles are used to represent sets, and shading is used to represent the portion we are describing. Consider the following Venn diagrams to represent $\overline{A}$, $A \cup B$, and $A \cap B$.

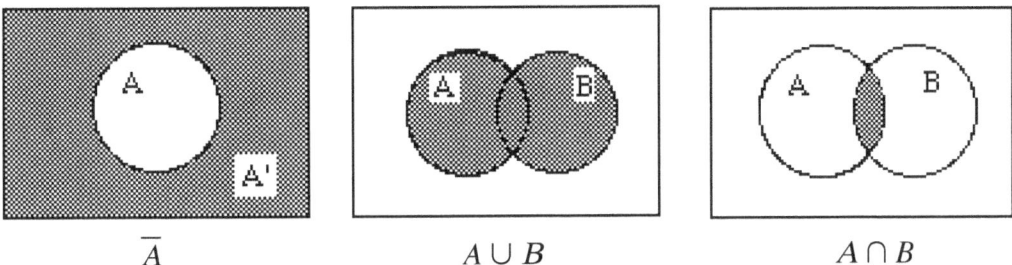

$\overline{A}$ $A \cup B$ $A \cap B$

Figure 7.2.2 Venn diagrams for basic set operations.

Venn diagrams are particularly useful to describe more complicated set operations. Consider the set $\overline{A \cup B}$. The union of the sets A and B is formed first, followed by taking the complement. We can use a sequence of Venn diagrams to illustrate these two operations.

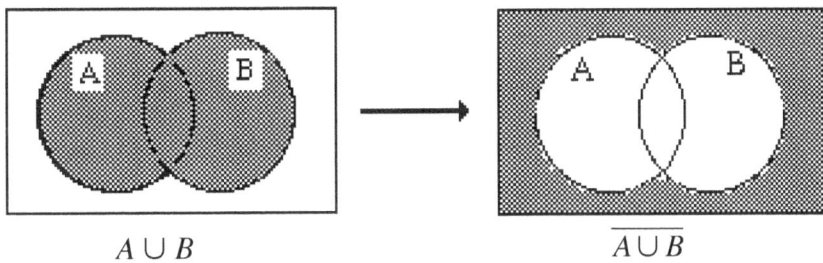

$A \cup B$ $\overline{A \cup B}$

Figure 7.2.3 Venn diagram for $\overline{A \cup B}$.

A representation for $\overline{A} \cup \overline{B}$ might be developed by the following sequence of Venn diagrams.

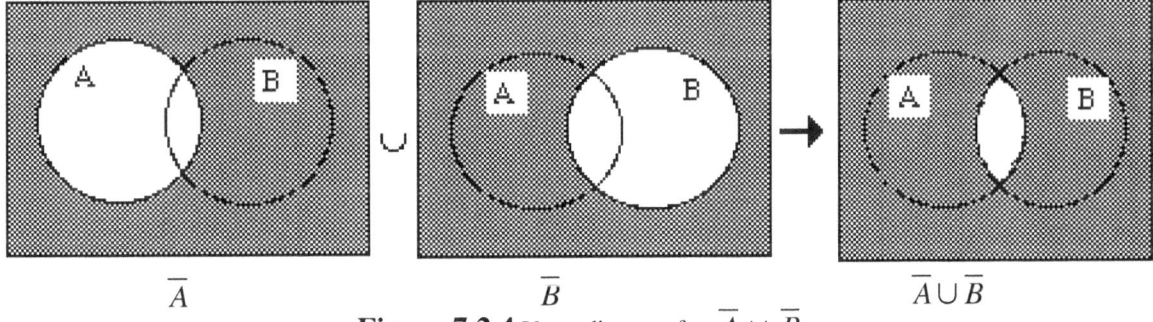

$\overline{A}$ $\overline{B}$ $\overline{A} \cup \overline{B}$

Figure 7.2.4 Venn diagram for $\overline{A} \cup \overline{B}$.

Comparing the diagrams for $\overline{A \cup B}$ and $\overline{A} \cup \overline{B}$ clearly shows, in general, that the two sets need not be equal. Consider the Venn diagram sequence for $\overline{A} \cap \overline{B}$ in Figure 7.2.5.

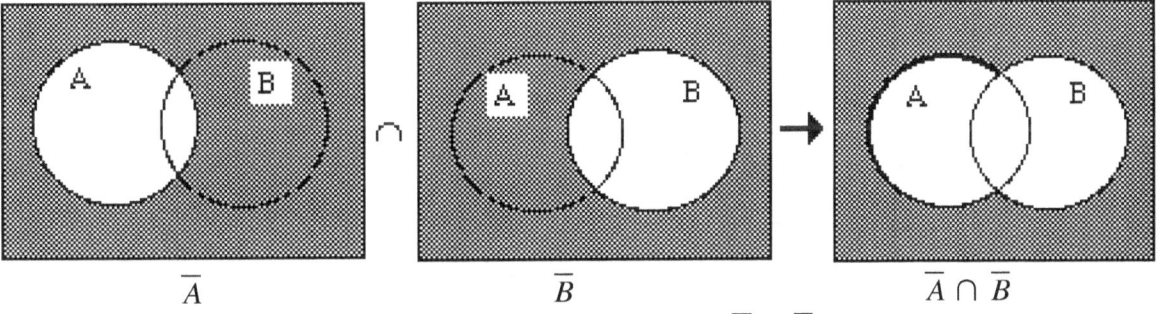

$\overline{A}$ $\overline{B}$ $\overline{A} \cap \overline{B}$

Figure 7.2.5 Venn diagram for $\overline{A} \cap \overline{B}$.

Notice that this diagram does agree with that for $\overline{A \cup B}$. In fact, this equality is part of a famous pair of set equalities known as ***De Morgan's Laws***. For any sets A and B:

$$\overline{A \cup B} = \overline{A} \cap \overline{B} \qquad \text{and} \qquad \overline{A \cap B} = \overline{A} \cup \overline{B}$$

Verification of the second law is left as an exercise. You may recall that we have already encountered the symbolic logic version of De Morgan's Laws in Exercise Set 3.3 of Chapter 3 on Logic. These properties are useful in simplifying more complicated set operations.

Augustus De Morgan (1806-1871), who was born in India to a British military family, suffered the loss of vision in his right eye while still an infant. This disability motivated teasing from his childhood classmates and perhaps influenced his development into a rather gruff and strongly opinionated individual. He entered Trinity College in Cambridge at the age of 16 and later became the first professor of mathematics at University College in London in 1828. He wrote several mathematics textbooks and numerous articles over his lifetime and was known as a great reformer of mathematical logic.

Figure 7.2.6 Augustus De Morgan (1806-1870)

Example 5
Let $U = \{1, 2, 3, 4, 5, 6, 7, 8, 9, 10\}$, $P = \{2, 3, 5, 7\}$, $E = \{2, 4, 6, 8, 10\}$. Find the set $\overline{P \cup \overline{E}}$ directly and by using De Morgan's Laws.

Solution
Directly, $P = \{2, 3, 5, 7\}$ and $\overline{E} = \{1, 3, 5, 7, 9\}$, so $P \cup \overline{E} = \{1, 2, 3, 5, 7, 9\}$, and thus $\overline{P \cup \overline{E}} = \{4, 6, 8, 10\}$. By De Morgan's Laws,

$$\begin{aligned} \overline{P \cup \overline{E}} \quad &= \overline{P} \cap E \quad \text{(since } \overline{\overline{E}} = E) \\ &= \{1, 4, 6, 8, 9, 10\} \cap \{2, 4, 6, 8, 10\} \\ &= \{4, 6, 8, 10\} \; \blacklozenge \end{aligned}$$

Notice that using De Morgan's Laws in the last example simplifies the amount of set operations necessary. Venn diagrams are also useful when counting is involved with set operations. The ***cardinality*** of a finite set A is the number of objects in the set and is denoted by $n(A)$. The concept of cardinality is used primarily in the study of infinite sets, which we will not

148

consider in this text. So, for our purposes, cardinality is simply a function whose domain is the subsets of a universal set and whose range is the set of non-negative integers (representing the number of elements in the set). Clearly, we see that $n(\varnothing) = 0$ and that if A and B are sets such that $A \subseteq B$, then $n(A) \leq n(B)$. We wish to develop rules for finding the cardinality of unions and intersections of sets whose cardinalities are known.

Example 6

If $A =$ {letters of the English alphabet} and $B = \{x \mid 1 \leq x \leq 100$ and x is an even integer}, then $n(A) = 26$ and $n(B) = 50$. Since $A \cap B = \varnothing$, we see that $n(A \cap B) = 0$ and $n(A \cup B) = 76$. ◆

Suppose we survey 100 students and find that 48 are currently taking an English course, 56 are currently taking a mathematics course, and 22 are taking both English and mathematics. How can we find out how many students are taking neither English nor mathematics? Let E represent the event (set) of taking an English course and M the event of taking a mathematics course. The following Venn diagram illustrates a partition of U (the universal set of 100 students surveyed) into *four* distinct sets. This is an extension of the definition of a partition of U in which the union of several sets equals U and no pair of the sets intersect.

$E \cap M =$ {students taking both English and mathematics}
$E \cap \overline{M} =$ {students taking English but not mathematics}
$\overline{E} \cap M =$ {students taking mathematics but not English}
$\overline{E} \cap \overline{M} =$ {students taking neither English nor mathematics}

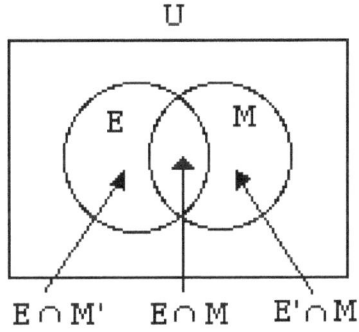

U

E ∩ M'   E ∩ M   E' ∩ M

Figure 7.2.7 Partition of U into four sets. ($\overline{E} = E'$, $\overline{M} = M'$)

We know there are 22 students taking both English and mathematics, so $n(E \cap M) = 22$. Now notice $E \cap M$ and $\overline{E} \cap M$ form a partition of event M. Therefore,

$$n(E \cap M) + n(\overline{E} \cap M) = n(M)$$

We also know there are 56 students taking mathematics, so $n(M) = 56$. Substituting these values into the above equation,

$$22 + n(\overline{E} \cap M) = 56$$
$$n(\overline{E} \cap M) = 34$$

Similarly, we partition event E into the two sets $E \cap M$ and $E \cap \overline{M}$, to obtain,

$$n(E \cap M) + n(E \cap \overline{M}) = n(E)$$
$$22 + n(E \cap \overline{M}) = 48$$
$$n(E \cap \overline{M}) = 26$$

Now we re-draw the Venn diagram with these computed values inserted.

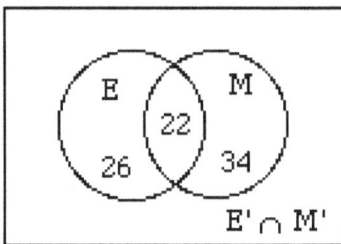

Figure 7.2.8 Partition of U into four sets. ($\overline{E} = E'$, $\overline{M} = M'$)

So we have

$$n(E \cap \overline{M}) + n(E \cap M) + n(\overline{E} \cap M) + n(\overline{E} \cap \overline{M}) = n(U)$$
$$26 + 22 + 34 + n(\overline{E} \cap \overline{M}) = 100$$
$$n(\overline{E} \cap \overline{M}) = 18$$

We see that 18 students surveyed were taking neither English nor mathematics. Note therefore that the number of students taking *either* English or mathematics must be
$$n(E \cup M) = 100 - 18 = 82.$$

Of course, this could also have been obtained by observing in Figure 7.2.8 that
$$n(E \cup M) = 26 + 22 + 34 = 82.$$

We emphasize that this is *not* equal to $n(E) + n(M) = 48 + 56 = 104$, since that would be counting the number of students in both sets, $n(E \cap M)$, twice. Note that if we subtract this value, 22, from 104, then of course, we would again arrive at the correct value of 82. This is shown by the equation

$$n(E \cup M) = n(E) + n(M) - n(E \cap M) = 48 + 56 - 22 = 82.$$

This a specific application of the property known as the ***Addition Principle***:

If A and B are sets, then $n(A \cup B) = n(A) + n(B) - n(A \cap B)$.

Reviewing this last example, we see that most of the work could have been done by drawing a single Venn diagram and then subtracting values to find partition elements.

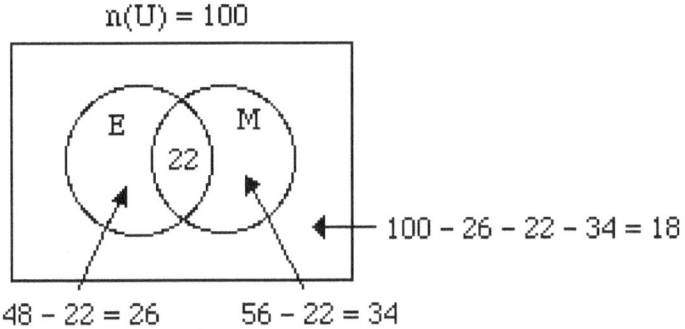

Figure 7.2.9 Simplified partition diagram for *U*.

So we see that partitions, Venn diagrams, and the Addition Principle are quite useful for working with counting problems of this type.

Example 6

Let $A = \{1, 3, 4, 5, 7, 8, 10\}$ and $B = \{3, 4, 7, 9, 10\}$. Verify the Addition Principle by first computing $A \cup B$ and $A \cap B$ directly.

Solution

$$A \cup B = \{1, 3, 4, 5, 7, 8, 9, 10\}, \text{ and so } n(A \cup B) = 8.$$
$$A \cap B = \{3, 4, 7, 10\}, \text{ and so } n(A \cap B) = 4.$$

Since $n(A) = 7$ and $n(B) = 5$, the formula $n(A \cup B) = n(A) + n(B) - n(A \cap B)$ is easily verified. ◆

Example 7

A universal set *U* contains 200 elements. If $n(A) = 85$, $n(\overline{B}) = 138$, and $n(A \cup B) = 107$, for sets *A* and *B*, then find: (i) $n(A \cap B)$, (ii) $n(A \cap \overline{B})$, and (iii) $n(\overline{A} \cap B)$.

Solution

(i) First we must have $n(B) = n(U) - n(\overline{B}) = 200 - 138 = 62$.
Then, since the Addition Principle implies that $n(A \cap B) = n(A) + n(B) - n(A \cup B)$, it must true that
$$n(A \cap B) = 85 + 62 - 107 = 40.$$
(ii) Since $A \cap B$ and $A \cap \overline{B}$ form a partition of *A*,
$$n(A \cap \overline{B}) = n(A) - n(A \cap B) = 85 - 40 = 45.$$

(iii) Since $A \cap B$ and $\overline{A} \cap B$ form a partition of *B*,
$$n(\overline{A} \cap B) = n(B) - n(A \cap B) = 62 - 40 = 22. ◆$$

151

Exercise Set 7.2

1. True or False. Let A and B be any subsets of the universal set U.

   i) $A \subseteq A \cap B$

   ii) $A \subseteq A \cup B$

   iii) $n(A) + n(\overline{A}) = n(U)$

   iv) $A \cap \varnothing = \varnothing$

   v) $\overline{A \cap B} = \overline{A} \cap \overline{B}$

   vi) If $A \subseteq B$, then $n(A) \leq n(B)$

   vii) If A and B are disjoint subsets, then $\overline{A} \cap \overline{B} = \varnothing$.

   viii) If A and B are disjoint subsets, then $n(A \cup B) = n(A) + n(B)$.

2. Let $U = \{$all college students$\}$, $M = \{$students currently taking mathematics$\}$, $H = \{$students currently taking history$\}$, and $B = \{$students currently taking biology$\}$. Describe (in words) each of the following sets:

     (i) $M \cap B$ (ii) $\overline{H}$ (iii) $\overline{H} \cap \overline{B}$ (iv) $M \cup H$

3. Let $U = \{$people who like sports$\}$, $B = \{$people who like basketball$\}$, $F = \{$people who like football$\}$, and $H = \{$people who like hockey$\}$. Describe (in words) each of the following sets:

     (i) $B \cup H$ (ii) $\overline{F}$ (iii) $\overline{B} \cap F$ (iv) $\overline{H \cup F}$

4. Let $U = \{5, 6, 7, 8, 9, 10, 11\}$, $A = \{6, 7, 8\}$, and $B = \{7, 9, 11\}$. List the elements of each of the following sets:

     (i) $\overline{A}$ (ii) $A \cup B$ (iii) $\overline{A} \cap B$ (iv) $\overline{A \cap B}$ (v) $\overline{A} \cup \overline{B}$

     (f) Do A and B form a partition of U?

5. Let $U = \{-3, -2, -1, 0, 1, 2, 3\}$, $A = \{-2, -1, 0, 1\}$, and $B = \{-1, 1, 2\}$. List the elements of each of the following sets:

     (i) $\overline{B}$ (ii) $A \cap B$ (iii) $\overline{A} \cup B$ (iv) $\overline{A \cup B}$ (v) $\overline{A \cup \overline{B}}$

6. Draw Venn diagrams illustrating each of the following sets:

     (i) $A \cap \overline{B}$ (ii) $\overline{A} \cup B$ (iii) $\overline{A \cup B}$ (iv) $\overline{A \cap \overline{B}}$

7. Draw Venn diagrams illustrating each of the following sets:

     (i) $\overline{A} \cap B$ (ii) $A \cup \overline{B}$ (iii) $\overline{\overline{A} \cap B}$ (iv) $\overline{\overline{A} \cup B}$

8. De Morgan's second law states that $\overline{A \cap B} = \overline{A} \cup \overline{B}$. Draw Venn diagrams for $\overline{A \cap B}$ and $\overline{A} \cup \overline{B}$ to verify these two sets are equal.

9. Let $U = \{1, 2, 3, 4, 5, 6, 7, 8\}$, $F = \{2, 4, 5, 7, 8\}$, and $S = \{3, 4, 8\}$. Answer each of the following questions:

i) Find the set $\overline{F \cap \overline{S}}$ directly and by using De Morgan's Laws.

ii) Find the set $\overline{\overline{F} \cup S}$ directly and by using De Morgan's Laws.

iii) Find the sets $F \cap S$ and $F \cap \overline{S}$, then verify that they form a partition of F.

iv) Find the sets $S \cap F$ and $S \cap \overline{F}$, then verify that they form a partition of S.

10. Let $U = \{a, b, c, d, e, f, g\}, P = \{a, d, f, g\}$, and $Q = \{b, c, d, g\}$. Answer each of the following questions:

i) Find the set $\overline{\overline{P} \cup Q}$ directly and by using De Morgan's Laws.

ii) Find the set $\overline{P \cap \overline{Q}}$ directly and by using De Morgan's Laws.

iii) Find the sets $P \cap Q$ and $P \cap \overline{Q}$, then verify that they form a partition of P.

iv) Find the sets $Q \cap P$ and $Q \cap \overline{P}$, then verify that they form a partition of Q.

11. Any group of people can be used as a universal set that can be partitioned by natural characteristics, such as sex, age, etc. Describe some other characteristics that could be used to partition the students in your mathematics class.

12. Describe a characteristic that could be used to partition the members of the U.S. Congress.

13. For any set A, explain why $\varnothing \subseteq A$ and $A \subseteq A$.

14. For any sets A and B, explain why $A \cap B \subseteq A$ and $A \cap B \subseteq B$.

15. The *set difference* $A - B$ is defined by the equation $A - B = A \cap \overline{B}$. Draw Venn diagrams illustrating each of the following sets:

i) $A - B$ ii) $B - A$ iii) $A - \overline{B}$ iv) $\overline{B - A}$

16. The set $A - B$ was defined in the previous exercise. Use a Venn diagram to illustrate that
$$n(A \cup B) = n(A - B) + n(B - A) + n(A \cap B).$$

17. If $n(A \cup B) = 50, n(A - B) = 21, n(B - A) = 10$, then find $n(A \cap B)$.

18. If $n(A \cup B) = 100, n(A - B) = 63, n(B - A) = 15$, then find $n(A)$.

19. If $n(U) = 125, n(A \cup B) = 100, n(A - B) = 40, n(B - A) = 53$, then find $n(A \cap B), n(B)$, and $n(\overline{B})$.

20. Let g be a function which assigns to each set A the number of subsets that can be formed from A. For example, $g(\{0, 1\}) = 4$, since the subsets of $\{0, 1\}$ are $\varnothing, \{0\}, \{1\}$, and $\{0, 1\}$. (Recall that $\varnothing$ and A are always subsets of A.) Find the following:

i) $g(\{1, 2, 3\})$ ii) $g(\{3\})$ iii) $g(\varnothing)$ iv) $g(\{a, b, c, d\})$

21. Based on your answers to the previous exercise, create a formula for $g(A)$ for any set A in terms of $n(A)$. How many subsets of cards exist for a deck of 52 cards?

22. Let $U = \{1, 2, 3, \ldots, 9, 10\}$. Define the set function f by the rule $f(A) = \overline{A}$ for any set A. Find the following:

   i) $f(\{1, 3, 5, 7, 9\})$ ii) $f(\{1, 4, 5, 9\})$ iii) $f(\varnothing)$ iv) $f(U)$

23. Let $U = \{\alpha, \beta, \chi, \delta, \varepsilon, \phi, \gamma\}$, $L = \{\alpha, \chi, \varepsilon, \phi\}$, and $M = \{\beta, \lambda\}$. Define the set function f by the rule $f(A) = \overline{A}$ for any set A. Find the following:

   i) $f(L)$ ii) $f(L \cap M)$ iii) $n(L)$ iv) $n(f(L))$ v) $n(f(L \cap M))$

24. Let U, L, M and f be defined as in the previous problem. Find the following:

   i) $f(M)$ ii) $n(M)$ iii)) $n(f(M))$ iv) $f(U)$ v) $n(f(U))$

25. In a survey of 1000 college students, 430 are currently taking chemistry, 640 are currently taking history, and 225 are currently taking both chemistry and history.

   i) How many students are currently taking either chemistry or history?
   ii) How many students are taking chemistry, but not history?
   iii) How many students are taking neither chemistry nor history?

26. A biologist examines 100 trees, and finds that 80 of them are either pine trees or are at least 20 feet high. She also finds that 60 of the trees are pine trees, and 55 of the trees are at least 20 feet high.

   i) How many trees are pine trees that are at least 20 feet high?
   ii) How many trees are not pine trees and are less than 20 feet high?
   iii) How many trees are at least 20 feet high but are not pine trees?

27. A sample of 600 vegetables were exposed to a certain level of pesticide PCH-2 for several months. They were then tested and 255 had become vitamin depleted, 338 mineral depleted, and 68 had lost both their vitamin and mineral content.

   i) How many vegetables were depleted in vitamins or minerals?
   ii) How many vegetables were depleted in minerals, but not in vitamins?
   iii) How many vegetables were unaffected by PCH-2?

28. Human blood can contain no antigens, the A antigen, the B antigen, or both the A and B antigens. Blood types are then characterized as:

   Type A: contains the A antigen but not the B antigen
   Type B: contains the B antigen but not the A antigen
   Type AB: contains both A and B antigens
   Type O: contains neither A nor B antigens

In 100 samples of blood, 40 are found to contain the A antigen, 30 are found to contain the B antigen, and 8 are found to contain both antigens.

   (i) How many of the samples are Type A blood?
   (ii) How many of the samples are Type B blood?
   (iii) How many of the samples are Type O blood?

29. Antigens were defined in the previous exercise. A third antigen, called the Rh antigen, may or may not be present in human blood. Blood which contains the Rh antigen is called "positive", thus blood containing all three antigens is called Type AB-positive, and blood which does not contain the *Rh* antigen is called "negative", thus blood containing no antigens is called Type *O*-negative. The following Venn diagram summarizes samples of blood:

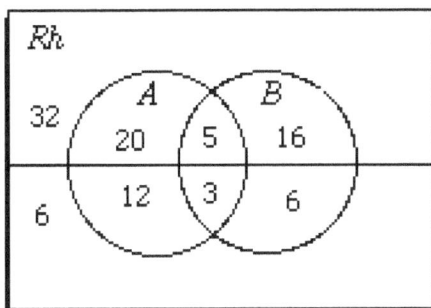

Figure 7.2.10 Blood antigens Venn diagram.

(i) How many samples are Type *O*-negative?
(ii) How many samples are Type *B*-positive?
(iii) How many samples are characterized as "positive"?
(iv) How many samples contain exactly two antigens?
(v) How many samples contain exactly one antigen?

30. Use DeMorgan's laws and the Addition Principle to show that
$$n(\overline{A \cap B}) = n(\overline{A}) + n(\overline{B}) - n(\overline{A \cup B}).$$

31. Venn diagrams can be used to display situations involving more than two sets. The stained glass window below was created by Maria McClafferty in honor of John Venn and is located in the dining hall of Gonville and Caius College in Cambridge, UK. Let *A* be the interior of the top circle, let *B* be the interior of the bottom leftmost circle and let *C* be the interior of the bottom rightmost circle.

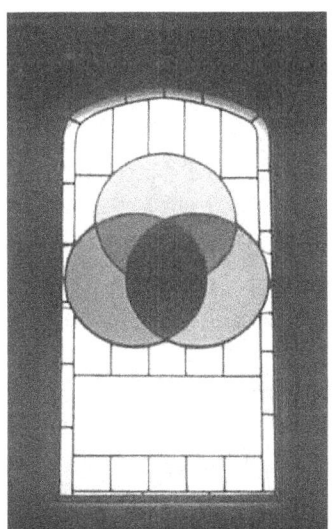

Use set operations to denote each of these regions:
i) The three-sided figure in the middle.
ii) The part of *A* not contained in *B* or *C*.
iii) The part of *B* not contained in *A* or *C*.
iv) The part of *A* or *B* not contained in *C*.
v) The part of *B* and *C* not contained in *A*.

7.3 Basic Probability

Now, the value of an idea has nothing whatsoever to do with the sincerity of the man who expresses it. Indeed, the probabilities are that the more insincere the man is, the more purely intellectual will the idea be...

Oscar Wilde

The Picture of Dorian Gray

The first systematic study of probability was probably done by a gambler. Observation of patterns is the essence of inspiration for mathematics and certainly any successful gambler is sure to be adept at identifying patterns that occur in card and dice games and using this knowledge to his advantage. An exceptional example of this is the renowned Italian scientist **Girolamo Cardano** (1501-1576). In spite of suffering the social obstacles imposed by an illegitimate birth, Cardano (or the Latin *Cardan*) rose to fame and prominence as one of the most brilliant physicians and mathematicians of 16th century Italy. At different times in his life, he was a university professor of mathematics and of medicine and even served as rector of the prestigious College of Physicians in Milan.

Figure 7.3.1 Girolamo Cardano

Unfortunately, his addiction to gambling led not only to much unhappiness in his personal life, but also robbed him of time that he could have used to produce even more contributions than the considerable amount he accomplished. One positive result of this affliction, however, was that he utilized his keen mathematical abilities to create one of the first formal analyses of the rules of chance and probability.

In examining random phenomena, we are not only interested in what can occur (the sample space), but also in measuring the chances that a particular event occurs. For instance, if you toss a fair coin, what are the chances that the face shows a head? Intuitively we might say

156

that the chances that a head appears to be $\frac{1}{2}$, or 50%. What does this mean? Put another way, if we toss the coin a large number of times, we might expect that heads would occur in about half the number of tosses. More explicitly, we are using the *relative frequency approach* to defining *probability*. The **relative frequency** of an event A occurring is given by:

$$\frac{\text{\# times } A \text{ occurs}}{\text{\# times experiment repeated}}$$

Coin Tosses	# Heads	Relative Frequency
50	31	31/50 = 0.62
100	54	54/100 = 0.54
250	118	118/250 = 0.472
500	263	263/500 = 0.526
1000	489	489/1000 = 0.489
5000	2531	2531/5000 = 0.506

Figure 7.3.2 Relative frequencies of tossing heads.

As seen in Figure 7.3.2, in the experiment of tossing a coin, the fraction of heads gets increasingly closer to $\frac{1}{2}$ as you keep repeating the experiment. It seems reasonable to call this limiting value the **probability** of tossing a heads and this method for assigning probabilities is referred to formally as the **Law of Large Numbers**.

If p represents the probability that an event A occurs, and r represents the relative frequency of A, then r approaches p as the number of repetitions increases.

Therefore, we assign probabilities to events (or outcomes) in such a way that the relative frequency agrees with our assignments in the long run. Next we need to develop a reasonable way to assign probabilities that satisfy the Law of Large Numbers. First, we define a **uniform sample space** to be a sample space for an experiment that has a finite number of outcomes each of which is equally likely to occur. For example, $S_1 = \{H, T\}$ is a uniform sample space for tossing a fair coin once. Similarly, $S_2 = \{HH, HT, TH, TT\}$ is a uniform sample space for tossing a fair coin twice (or tossing two fair coins). It is important to distinguish S_2 from $S_3 = \{0 \text{ heads}, 1 \text{ head}, 2 \text{ heads}\}$, which is also a sample space for tossing a fair coin twice, because S_3 is not a *uniform* sample space. As you can see comparing S_2 and S_3, the outcome "1 head" will occur twice as often as the outcomes "0 heads" or "2 heads", thus S_3 does not contain equally likely outcomes.

We can now make the following definition. If S is a uniform sample space and A is an event of S, then the ***probability of event*** A, denoted $p(A)$, is the *function* defined by

$$p(A) = \frac{n(A)}{n(S)} \; .$$

It is important to point out that probability is a function whose domain is all events (subsets) of a sample space and whose range is the interval $[0, 1]$. Note that $p(A) = 0$ must mean that A contains no elements (meaning event A *cannot* occur), and so A must be the empty set $\varnothing$. Further, $p(A) = 1$ must mean that $n(A) = n(S)$, so A contains all elements of the sample space (meaning that A *must be* S).

Example 1

A fair coin is tossed twice. Then $S = \{HH, HT, TH, TT\}$ can be used to represent a uniform sample space for this experiment. Given events A, B, C listed as:

$\quad\quad$ A: at least 1 head occurs $= \{HH, HT, TH\}$
$\quad\quad$ B: exactly 1 head occurs $= \{HT, TH\}$
$\quad\quad$ C: more than two heads occur $= \varnothing$

Then the probabilities of A, B, and C are given as:

$$p(A) = \frac{n(A)}{n(S)} = \frac{3}{4}$$

$$p(B) = \frac{n(B)}{n(S)} = \frac{2}{4} = \frac{1}{2}$$

$$p(C) = \frac{n(C)}{n(S)} = \frac{0}{4} = 0 \; \blacklozenge$$

The next example may be of interest to those of you with an interest in the gambling game of "craps" played at all casinos.

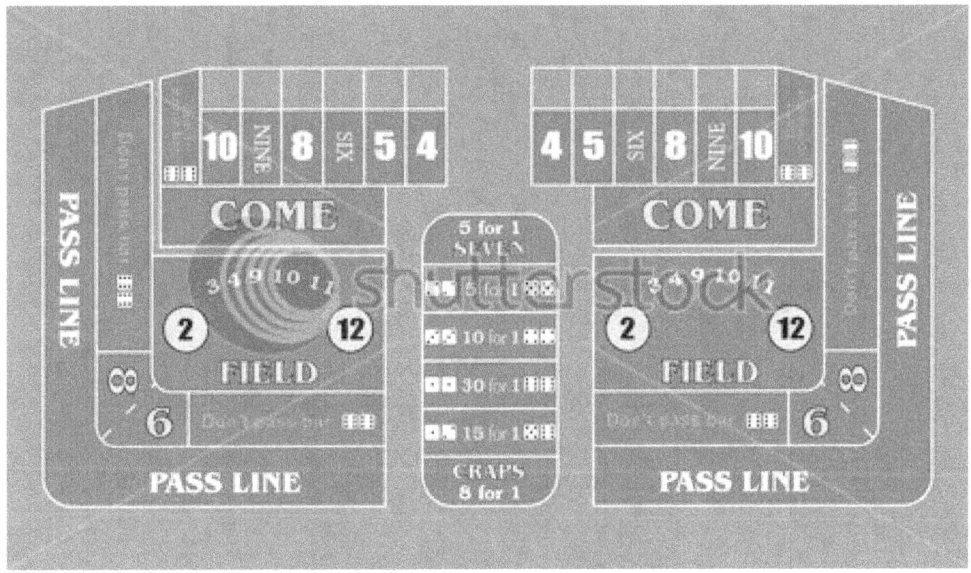

A typical "Craps" gaming table.

Example 2

Two fair dice are rolled and their outcomes are recorded. Then S can be listed as in Figure 7.3.3.

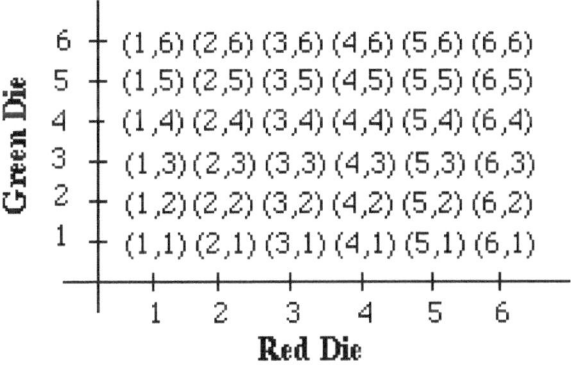

Figure 7.3.3 Two dice chart.

Note that S is a uniform sample space. Find the probabilities of the following events.

    (i) Doubles are rolled.
    (ii) The sum of the dice is greater than 8.
    (iii) The sum of the dice is even and the first die is odd.

Solution

    (i) Let D = "Doubles" = $\{(1,1), (2,2), (3,3), (4,4), (5,5), (6,6)\}$.
Then $n(D) = 6$ while $n(S) = 36$. So,

$$p(D) = \frac{n(D)}{n(S)} = \frac{6}{36} = \frac{1}{6}$$

Note that we can also list D graphically, where doubles are underlined.

159

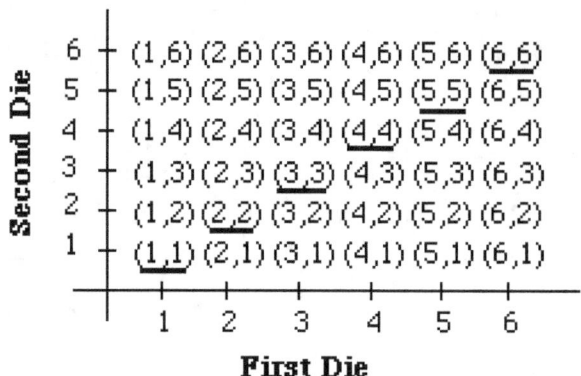

Figure 7.3.4 Two dice chart for doubles.

(ii) Let G = "The sum of the dice is greater than 8". The sum must therefore be 9, 10, 11, or 12. These pairs are underlined in Figure 7.3.5.

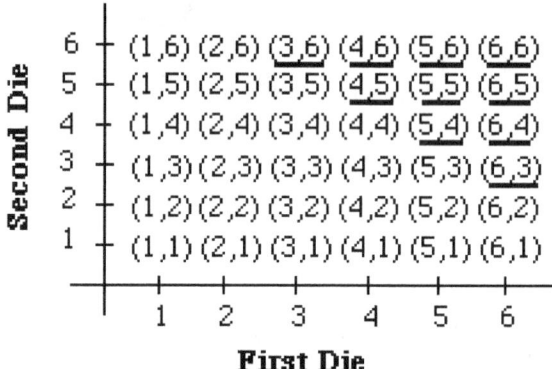

Figure 7.3.5 Two dice chart for sum greater than 8.

Then $n(G) = 10$ while $n(S) = 36$. So, $p(G) = \dfrac{n(G)}{n(S)} = \dfrac{10}{36} = \dfrac{5}{18}$.

(iii) Let B = "The sum of the dice is even and the first die is odd". The possible die combinations are underlined in Figure 7.3.6.

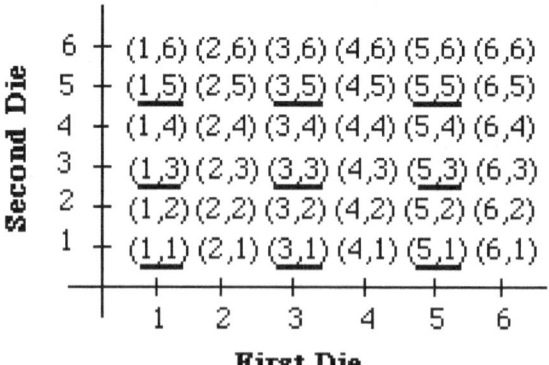

Figure 7.3.6 Two dice chart for part (iii).

Then $n(B) = 9$ while $n(S) = 36$, so $p(B) = \dfrac{n(B)}{n(S)} = \dfrac{9}{36} = \dfrac{1}{4}$ ♦

It is important to note that we are not considering the probability of events in which the sample space is not uniform.

In summary, we have defined probability as a function whose domain is all subsets A of a uniform sample space and whose range is the interval $[0, 1]$. This functional assignment is determined by the formula,

$$p(A) = \frac{n(A)}{n(S)} \ .$$

Thus the assignment is really a matter of counting outcomes in our sample space (assuming it is finite). We now consider some very useful properties of the probability function.

Properties of Probability

If S is a uniform sample space which is finite and non-empty, and A and B are events, the following properties hold.

1. $p(S) = 1$
2. $p(\varnothing) = 0$
3. $0 \le p(A) \le 1$
4. $p(x) = \dfrac{1}{n(S)}$ for any outcome $x \in S$
5. $p(A \cup B) = p(A) + p(B) - p(A \cap B)$
6. $p(A \cup B) = p(A) + p(B)$ if A and B are mutually exclusive
7. $p(\overline{A}) = 1 - p(A)$, where $\overline{A}$ is the complement of A

We will verify some of these properties.

1. $p(S) = 1$ directly from the definition $p(A) = \dfrac{n(A)}{n(S)} \ .$

$$p(S) = \frac{n(S)}{n(S)} = 1$$

Recall the $p(A)$ is a *function*, so $p(S)$ simply means to evaluate that function at $A = S$.

2. $p(\varnothing) = 0$ directly from the definition $p(A) = \dfrac{n(A)}{n(S)} \ .$

$$p(\varnothing) = \frac{n(\varnothing)}{n(S)} = \frac{0}{n(S)} = 0$$

3. Recall that A is an event (hence a subset of S), so $\varnothing \subseteq A$ and $A \subseteq S$. Then $n(\varnothing) \le n(A)$ and $n(A) \le n(S)$. Therefore, $0 \le n(A) \le n(S)$. Dividing by $n(S)$,

$$\frac{0}{n(S)} \le \frac{n(A)}{n(S)} \le \frac{n(S)}{n(S)}$$
$$0 \le p(A) \le 1$$

4. This proof is left as an exercise.

5. This property should look similar to the Addition Principle.
$$n(A \cup B) = n(A) + n(B) - n(A \cap B)$$

Dividing each term by $n(S)$,
$$\frac{n(A \cup B)}{n(S)} = \frac{n(A)}{n(S)} + \frac{n(B)}{n(S)} - \frac{n(A \cap B)}{n(S)}$$
$$p(A \cup B) = p(A) + p(B) - p(A \cap B)$$

The proofs of properties 6 and 7 are left as exercises.

The following example illustrates some of these properties.

Example 3

Suppose A and B are events such that $p(A) = \frac{1}{2}$, $p(B) = \frac{1}{3}$, and $p(A \cap B) = \frac{1}{6}$. Find the following probabilities:

(i) $p(\overline{A})$ (ii) $p(\overline{B})$ (iii) $p(A \cup B)$ (iv) $p(\overline{A} \cup \overline{B})$ (v) $p(\overline{A} \cap \overline{B})$

Solution

(i) Using Property 7, we have,
$$p(\overline{A}) = 1 - p(A) = 1 - \frac{1}{2} = \frac{1}{2}$$

(ii) Using Property 7, we have,
$$p(\overline{B}) = 1 - p(B) = 1 - \frac{1}{3} = \frac{2}{3}$$

(iii) Using Property 5, we have,
$$p(A \cup B) = p(A) + p(B) - p(A \cap B) = \frac{1}{2} + \frac{1}{3} - \frac{1}{6} = \frac{2}{3}$$

(iv) Applying De Morgan's law, $\overline{A} \cup \overline{B} = \overline{A \cap B}$, we use Property 7,
$$p(\overline{A} \cup \overline{B}) = p(\overline{A \cap B}) = 1 - p(A \cap B) = 1 - \frac{1}{6} = \frac{5}{6}$$

(v) Applying De Morgan's law, $\overline{A} \cap \overline{B} = \overline{A \cup B}$, we use Property 7 and our result from part (iii).
$$p(\overline{A} \cap \overline{B}) = p(\overline{A \cup B}) = 1 - p(A \cup B) = 1 - \frac{2}{3} = \frac{1}{3} \quad \blacklozenge$$

Example 4

Let $S = \{1, 2, 3, 4, 5, 6, 7, 8\}$, $A = \{1, 3, 4, 5\}$, $B = \{2, 4, 6\}$, and $C = \{1, 3, 5, 8\}$. Find each of the following probabilities.

(i) $p(A)$ and $p(B)$ (ii) $p(A \cap C)$

(iii) $p(A \cup C)$ (iv) $p(\overline{B})$

(v) $p(\overline{A} \cap \overline{C})$ (vi) $p(B \cup C)$

Solution

(i) Since $n(A) = 4$, $n(B) = 3$, and $n(S) = 8$, we have,
$$p(A) = \frac{n(A)}{n(S)} = \frac{4}{8} = \frac{1}{2} \text{ and } p(B) = \frac{n(B)}{n(S)} = \frac{3}{8}$$

(ii) First we find $A \cap C = \{1, 3, 5\}$. So $n(A \cap C) = 3$ and
$$p(A \cap C) = \frac{n(A \cup C)}{n(S)} = \frac{3}{8}$$

(iii) Since $p(A \cap C) = \frac{3}{8}$, $p(A) = \frac{1}{2}$, and $p(C) = \frac{n(C)}{n(S)} = \frac{4}{8} = \frac{1}{2}$, we can use Property 5.
$$p(A \cup C) = p(A) + p(C) - p(A \cap C) = \frac{1}{2} + \frac{1}{2} - \frac{3}{8} = \frac{5}{8}$$

(iv) Since $\overline{B}$ is the complement of B, and $p(B) = \frac{3}{8}$, we can use Property 7.
$$p(\overline{B}) = 1 - p(B) = 1 - \frac{3}{8} = \frac{5}{8}$$

(v) There are a couple approaches we can use for this problem. We can do the problem directly by finding the required sets.
$$\overline{A} = \{2, 6, 7, 8\} \qquad \overline{C} = \{2, 4, 6, 7\} \qquad \overline{A} \cap \overline{C} = \{2, 6, 7\}$$

So the required probability is given by:
$$p(\overline{A} \cap \overline{C}) = \frac{n(\overline{A} \cap \overline{C})}{n(S)} = \frac{3}{8}$$

A quicker approach is to use De Morgan's laws, which state that
$\overline{A} \cap \overline{C} = \overline{A \cup C}$. Since $p(A \cup C) = \frac{5}{8}$, then using Property 7,
$$p(\overline{A \cup C}) = 1 - p(A \cup C) = 1 - \frac{5}{8} = \frac{3}{8}$$

(vi) Since $B \cap C = \varnothing$, then B and C are mutually exclusive, by Property 6 we have
$$p(B \cup C) = p(B) + p(C) = \frac{3}{8} + \frac{1}{2} = \frac{7}{8} \quad \blacklozenge$$

One application of probability that is useful in making decisions about financial risk-taking such as buying a lottery ticket or gambling in a casino concerns making predictions about your average return. More generally, if each outcome of an experiment is a numerical quantity (such as the payoff from a wager), then the ***expected value*** of the experiment is defined as the average value per occurrence of the experiment after a large number of repetitions. If X represents this numerical quantity, then the expected value of X is usually denoted by $E(X)$. For instance, suppose your friend offers to pay you one dollar every time you randomly draw a heart from a deck of bridge cards but collect 50 cents from you if you draw a club, spade, or diamond. If this is repeated indefinitely, you would win a dollar one-fourth of the time but lose 50 cents three-fourths of the time. In the long run you would expect a return of

$$E(X) = (+1)\left(\frac{1}{4}\right) + (-0.50)\left(\frac{3}{4}\right) = -0.125 \text{ dollars per repetition of this game. In general, if } X \text{ may}$$

be any one of a finite number of numerical values $x_1, x_2, \ldots, x_n$ associated with n outcomes of an experiment, then the expected value is defined to be

$$E(X) = x_1 p(x_1) + x_2 p(x_2) + \ldots + x_n p(x_n).$$

Example 5

Your school sponsors a raffle in which 1000 tickets are sold for a dollar each. Three tickets will be randomly drawn. One ticket number will win $100, the second ticket will win $50, and the third will win $25. If you buy one ticket, what is the expected value of your investment?

Solution

The probability that you win any single one of the prizes is 1/1000. Therefore, the expected value is

$$E(X) = (+99)\left(\frac{1}{1000}\right) + (+49)\left(\frac{1}{1000}\right) + (+24)\left(\frac{1}{1000}\right) + (-1)\left(\frac{997}{1000}\right) = -0.825 \, .$$

This means that, if this raffle was held many times and you bought one ticket each time, the average amount you would win is a negative –$0.825 per purchase. In other words you could expect to lose an average amount of 82.5 cents per raffle. Of course, most people enter raffles for the fun of it and to support a good cause. Just don't expect to get rich! ◆

We will conclude this brief look at probability by looking at an application in genetics discovered by Gregor Mendel (1822-1884) while cross-pollinating garden pea plants. He found that the flower color of a plant could be reasonably predicted by the flower color of its "parents" if he knew the genes responsible for the flower color. He discovered that each gene consists of a pair of *alleles* one of which is dominant and one which is recessive. For the gene determining color we let R represent the dominant red allele and let w represent recessive white allele. Then the possible combinations of alleles (called *genotypes*) are:

RR - red flower
Rw - red flower (since R is dominant and w is recessive)
wR - red flower (since R is dominant and w is recessive)
ww - white flower

In one experiment, Mendel crossed pure reds (*RR*) with pure whites (*ww*), and obtained all red-flowered peas. He then crossed these offspring, and obtained 705 red-flowered and 224 white-flowered offspring. If R is dominant and w is recessive, how can we explain these results? We

164

can describe the allele mixing by using a chart, called a Punnett square. If we cross pure reds (*RR*) with pure whites (*ww*), the Punnett square would look like:

	w	w
R	Rw	Rw
R	Rw	Rw

Note that all offspring are *Rw*, which are red-flowered (remember *R* is dominant and *w* is recessive). This is consistent with the first stage of Mendel's experiment. If we now cross these offspring, the Punnett square would look like:

	R	w
R	RR	Rw
w	wR	ww

Considering this last set of genotypes as a sample space, we have
$S = \{RR, Rw, wR, ww\}$. Since *RR*, *Rw*, and *wR* all produce red-flowered peas while *ww* produces white-flowered peas, we have the following probabilities.

$$p(\text{red-flowered}) = \frac{3}{4} = 0.75$$

$$p(\text{white-flowered}) = \frac{1}{4} = 0.25$$

To compare with Mendel's results on the $705 + 224 = 929$ plants, let's compute the relative frequencies of each color.

$$r(\text{red-flowered}) = \frac{705}{929} \approx 0.76$$

$$r(\text{white-flowered}) = \frac{224}{929} \approx 0.24$$

Our probabilities are consistent with the relative frequencies obtained by Mendel. This approach used in genetics extends to more complicated genetic traits, and can be used by genetic counselors when advising prospective parents who may carry genes for some human diseases.

Exercise Set 7.3

1. A fair coin is tossed twice and the coin faces are recorded.
     i) List the sample space S.
     ii) Find the probability that at most 1 head occurs.
     iii) Find the probability that more than 1 tail occurs.
     iv) Find the probability that the two coin faces are the same.

2. A fair coin is tossed three times and the coin faces are recorded.
     i) List the sample space S.
     ii) Find the probability that at least 2 heads occur.
     iii) Find the probability that no more than 1 head occurs.
     iv) Find the probability that more than 3 tails occur.

3. Two fair dice (six-sided) are rolled and their outcomes are recorded.
     i) List the sample space S.
     ii) Find the probability that the sum of pips is at most 6.
     iii) Find the probability that the sum of pips is odd and the first die is at least 5.
     iv) Find the probability that the first die is 2 less than the second.
     v) Find the probability that the first die is at least 2 less than the second.

4. Explain why Property 4 (under **Properties of Probability**) is true. (*Hint:* How many outcomes are there in S?)

5. Prove Property 6 (under **Properties of Probability**).

6. Prove Property 7 (under **Properties of Probability**).

7. A positive integer is picked at random between the numbers 10 and 20, inclusive, and its value is recorded.
     i) List the sample space S.
     ii) Find the probability that the number is even.
     iii) Without listing the event, find the probability that the number is odd.
     iv) Find the probability that the number is prime.
     Hint: Recall that a prime number is only divisible by 1 and itself.
     v) Find the probability that the number is a solution to the inequality $x^2 < 300$.

8. In the previous exercise, let $A = \{10, 12, 13, 15, 16\}$ and $B = \{12, 13, 14, 17, 18, 20\}$. Find each of the following probabilities.

i)	$p(A)$ and $p(B)$	ii)	$p(\overline{A})$ and $p(\overline{B})$
iii)	$p(A \cap B)$ and $p(A \cup B)$	iv)	$p(\overline{A} \cap \overline{B})$ and $p(\overline{A} \cup \overline{B})$
v)	$p(\overline{A} \cap B)$ and $p(\overline{A} \cup B)$	vi)	$p(A \cap \overline{B})$ and $p(A \cup \overline{B})$

9. Let A and B be events such that $p(A) = \frac{1}{2}$, $p(B) = \frac{3}{4}$, and $p(A \cap B) = \frac{3}{8}$. Find each of the following probabilities.

    i) $p(\overline{A})$ ii) $p(\overline{B})$ iii) $p(A \cup B)$ iv) $p(\overline{A} \cap \overline{B})$ v) $p(\overline{A} \cup \overline{B})$

10. Let A and B be events such that $p(A) = \frac{2}{3}$, $p(B) = \frac{1}{4}$, and $p(A \cup B) = \frac{5}{6}$. Find each of the following probabilities.

    i) $p(\overline{A})$ ii) $p(\overline{B})$ iii) $p(A \cap B)$ iv) $p(\overline{A} \cap \overline{B})$ v) $p(\overline{A} \cup \overline{B})$

11. Sets A, B, and C form a partition of U if $U = A \cup B \cup C$, $A \cap B = \varnothing$, $A \cap C = \varnothing$, and $B \cap C = \varnothing$. In this case, show that $p(A) + p(B) + p(C) = 1$.

12. A single card is picked at random from a deck of 52 cards. Find the probability that the card is:

    i) a king   ii) red   iii) a spade   iv) a king or a spade   v) a black ace

    vi) a face card (*Note*: Face cards are jacks, queens, and kings.)

13. A four-digit number is picked at random. (The first digit must be non-zero.) Find each of the following probabilities.

    i) The first digit is a 5.
    ii) The number is a multiple of 100.
    iii) The number is even.
    iv) The number is not a multiple of 10. (*Hint*: Find the probability that the number *is* a multiple of 10, then apply a property of probability.)

14. A fair coin is tossed four times and the coin faces are recorded. Find each of the following probabilities.

    (i)   Two heads occur. (*Note*: Recall this means *exactly* two heads occur.)
    (ii)   More than two heads occur.
    (iii) At most two heads occur.
    (iv) At least one head occurs.

15. One red die and one green die are rolled and their outcomes are recorded. (Both dice are four-sided with 1, 2, 3, and 4 pips on each side.) Find the probabilities of the following events.

   i) Both dice show an odd number of pips.
   ii) At least one die is odd.
   iii) The sum of the pips is less than 7.
   iv) The number of pips showing on the red die is larger than the green die.

16. A liberal arts mathematics class has the following composition of students.

	male	female
freshmen	2	3
sophomores	3	8
juniors	1	4
seniors	5	6

If a student is picked at random, find each of the following probabilities.
    i) The student is male.
    ii) The student is a junior.
    iii) The student is not a senior.
    iv) The student is a female upper-classman (i.e. junior or senior).

17. In the previous exercise, find the following probabilities.
    i) The student is female.
    ii) The student is not a junior.
    iii) The student is a male freshman or male senior.
    iv) The student is a male freshman or female junior.

18. Let $U = \{-3, -2, -1, 0, 1, 2, 3\}$, $A = \{-2, -1, 0, 1\}$, and $B = \{-1, 1, 2\}$. Find the probability of each of the following events, assuming a number is picked at random from U.

    i) $A \cap \overline{B}$ ii) $\overline{A} \cup B$ iii) $\overline{\overline{A} \cap B}$ iv) $\overline{A \cup \overline{B}}$

19. In a survey of 1000 college students, 430 are currently taking chemistry, 640 are currently taking history, and 225 are currently taking both chemistry and history. (See exercise #25 from Section 7.2.) What is the probability that a student chosen from this survey:
    i) is taking either chemistry or history?
    ii) is taking chemistry, but not history?
    iii) is taking neither chemistry nor history?

20. A biologist examines 100 trees, and finds that 80 of them are either pine trees or are at least 20 feet high. She also finds that 60 of the trees are pine trees, and 55 of the trees are at least 20 feet high. (See exercise #26 from Section 7.2.) What is probability that a tree in this study:
    i) is at least 20 feet high?
    ii) is less than 20 feet high?
    iii) is at least 20 feet high but not a pine?

21. A roulette wheel used in a casino has 38 slots numbered 1, 2, 3, ..., 36, as well as 0 and 00. The odd numbers are colored black, the even numbers are colored red, and the numbers 0 and 00 are colored green. A ball is rolled around the wheel and randomly drops into one of the slots. Find each of the following probabilities.

    i) The ball lands in a red slot.
    ii) The ball lands in a green slot.
    iii) The ball lands on a prime number.

iv) The ball lands in a slot numbered 18-25.

v) The ball lands in an even slot or into a green slot.

22. In the previous problem, suppose a $1 bet on a red slot returns $1 (plus the dollar you bet) if the ball lands on red. What is the expected value of such a bet?

23. Each week the state lottery pays out $100,000 to the person who has chosen the same 6-digit number that is randomly drawn from a barrel with one million such numbers in it. If each lottery ticket (with one number on it) costs $2, what is your expected value if you purchase one ticket?

24. Suppose you work in an office where there is a weekly football pool. For $5 you get to pick one digit. If your pick is the last digit in the aggregate total of points scored in professional football games that week, then you win $45. What is the expected value of this bet?

25. Suppose we cross-pollinate mixed-red peas (Rw) with white peas (ww). Construct a Punnett square and compute the probability of obtaining red-colored peas.

26. If we then cross-pollinated the red-colored offspring from the previous exercise, what percent of *their* offspring would you expect to be white?

27. Human hair color (brown and blonde) can be examined with Punnett squares, where brown (*B*) is dominant and blonde (*b*) is recessive. If two brown-haired people have a child, is it possible for that child to be blonde? In this case, what probability would you associate with the child having blonde hair?

28. In humans, left-handedness appears to be recessive to the dominant right-handedness. Suppose a right-handed person *A* has a parent who is left-handed.

i) If *A* mates with a left-handed person, what is the probability they produce a left-handed child?

ii) If *A* mates with a right-handed person who has a left-handed parent, what is the probability they produce a left-handed child?

iii) If *A* mates with a right-handed person, both of whose parents are right-handed, is it possible for them to have a left-handed child?

29. A type of dwarfism (characterized by short limbs) is controlled by a dominant allele. Is it possible for two dwarf parents to produce a normally proportioned child? In this case, what probability would you associate with the child being normally proportioned?

30. In the game of backgammon, you can "bump" your opponent based on the roll of two dice. If your opponent leaves a piece unprotected eight spots from you, and your dice total 8 (e.g. (5, 3)), then you can move to his position and "bump" him to the beginning of the board. If his unprotected piece is five spots from you, your dice can either total 5 (e.g. (3, 2)), or either of the two dice can be a 5 (e.g. (4, 5)), in order to "bump" him. Find the probability that you can "bump" your opponent based on the following situations.

i) His unprotected piece is 7 spots from you.
ii) His unprotected piece is 5 spots from you.
iii) He has two unprotected pieces 4 and 8 spots from you.
Note: You are finding the probability that either (or both) can be bumped.

31. (Continuation of previous exercise) If your opponent leaves one piece unprotected, how many spots from you will have the highest probability that you can bump him? (*Hint:* Bumps can occur from positions 1-12, inclusive. You will need to compute the probability for each position to find the highest value.)

32. Related to the idea of probability is that of odds. Whereas the probability of an event is

$$p(E) = \frac{n(E)}{n(S)} = \frac{\text{number of favorable outcomes}}{\text{number of total outcomes}},$$

we define the ***odds for an event*** as the ratio $p(E)/p(\overline{E})$. This, in turn, is then equal to,

$$\frac{p(E)}{1 - p(E)} = \frac{n(E)/n(S)}{1 - n(E)/n(S)}$$
$$= \frac{n(E)}{n(S) - n(E)}$$
$$= \frac{\text{number of favorable outcomes}}{\text{number of unfavorable outcomes}}.$$

For example, if a fair coin is tossed twice, we can list the outcomes as {HH, HT, TH, TT}, so the odds of obtaining at least one head are $\frac{3 \text{ favorable}}{1 \text{ unfavorable}}$, or 3 to 1. This is usually denoted 3 : 1, and it can be interpreted that three times as many outcomes have at least one head than no heads. Find the odds for the following events and experiments.

i) Experiment: Toss a fair coin 3 times. Event: Obtain at least 2 heads.
ii) Experiment: Roll two fair die. Event: Sum of die is at most 5.
iii) Experiment: Pick a person at random from a group of 4, including Hillary.
Event: Hillary is picked.
iv) Experiment: Pick a card at random from a standard 52-card deck.
Event: The card is red. *Note:* This situation is called "even odds."

33. In horse racing, odds are usually posted *against* a horse. The ***odds against an event*** are defined as the reciprocal of the odds in favor, i.e. $p(\overline{E})/p(E)$ for an event E. This is then the reciprocal of all the ratios listed in the above exercise. In horse racing, odds of 5 : 2 against the horse winning would mean that for every $5 bet against the horse, $2 is bet in favor of the horse. List the odds (reduced form) against the horse winning for the following betting amounts.

i) $20,000 against; $6,000 in favor
ii) $5,000 against; $5,000 in favor
iii) $2,000 against; $7,000 in favor

34. If the odds for an event E are $4 : 1$, what is $p(E)$?

35. If the odds against an event A are $3 : 7$, what is $p(A)$?

Chapter Glossary

cardinality   The cardinality $n(S)$ of a finite set S is the number of elements contained in S.

complement   The complement   of a set A consists of the elements in the universal set that are not elements of A.

De Morgan's Laws For any sets A and B, $\overline{A \cup B} = \overline{A} \cap \overline{B}$ and $\overline{A \cap B} = \overline{A} \cup \overline{B}$.

disjoint sets   Two sets whose intersection is empty i.e. two sets A and B for which $A \cap B = \varnothing$.

empty set   The set containing no elements. It is denoted by the symbol $\varnothing$.

event   A subset of outcomes from a sample space.

expected value The average value per repetition of a numerically valued experiment after a large number of repetitions.

experiment   An action or process for which the outcome cannot be predicted with certainty.

intersection of two sets The intersection $A \cap B$ of two sets A and B is the set of elements belonging to both A and B.

Law of Large Numbers If p represents the probability that an event A occurs, and r represents the relative frequency of A, then r approaches p as the amount of repetitions increases.

mutually exclusive events Events that cannot occur at the same time.

odds for an event The ratio of the probability that the event does occur to the probability that it does not occur.

odds against an event The ratio of the probability that the event does not occur to the probability that it does occur.

partition of a set S   A pair of subsets A and B such that $A \cap B = \varnothing$ and $A \cup B = S$.

probability of an event A (in a uniform sample space S) A function p with domain consisting of the subsets (events) of S and range consisting of the unit interval $[0, 1]$. For any event A,
$$p(A) = \frac{n(A)}{n(S)} .$$

relative frequency For an event A it is defined by: $\dfrac{\# \text{ times } A \text{ occurs}}{\# \text{ times experiment repeated}}$

sample space The set of all distinct possible outcomes of an experiment.

set A collection of objects.

subset Set A is a subset of set B if every element of A is also an element of B.

uniform sample space A finite sample space in which each outcome is equally likely to occur.

union of two sets The union $A \cup B$ of two sets A and B is the set of objects belonging to either A or B.

universal set A set of all possible elements associated with a certain phenomenon or experiment.

Venn diagram Picture of sets that displays a certain relationship among them.

Chapter Review Test

. True/False. Let A and B be subsets.

i) $A \cap B \subseteq A \cup B$

ii) If A is a subset of the universal set U, then $A \cup \overline{A} = U$.

iii) $A \subseteq A \cap B$

iv) $\overline{A \cap B} = \overline{A} \cap \overline{B}$

v) If A and B are mutually exclusive events, then $p(A \cap B) = 0$.

vi) If U has 5 elements in it, then the total number of subsets of U is 10.

vii) $A = (A - B) \cup (A \cap B)$

viii) If $A \subseteq B$, then $n(B) \leq n(A)$

2. Consider the experiment of throwing a six-sided die twice and recording the sum of the numbers that occur. Is $S = \{0, 1, 2, 3, 4, 5, 6\}$ a valid sample space? Why or why not?

3. Let $U = \{s, t, u, v, w, x, y, z\}$, $A = \{s, u, v, y\}$, and $B = \{t, u, w, x, y\}$. Determine each of the following:

    i) $\overline{A}$ ii) $A \cap B$ iii) $A \cap \overline{B}$ iv) $\overline{A} \cup B$

    v) $n(B)$ vi) $n(\overline{A} \cap B)$ vii) $n(\overline{A \cup B})$

4. Draw Venn diagrams illustrating each of the following sets:

            $\overline{A \cap B}$ $(A - B) \cup (B - A)$

5. If $n(A \cup B) = 205$, $n(A) = 143$, $n(B) = 87$, then find $n(A \cap B)$.

6. If $n(A \cup B) = 400$, $n(A - B) = 229$, $n(B - A) = 150$, then find $n(A)$.

7. Let A and B be events such that $p(A) = \frac{1}{2}$, $p(B) = \frac{3}{4}$, and $p(A \cap B) = \frac{3}{8}$. Find the following probabilities.

    i) $p(\overline{A})$ ii) $p(\overline{B})$ iii) $p(A \cup B)$ iv) $p(\overline{A} \cap \overline{B})$ v) $p(\overline{A} \cup \overline{B})$

  8. In a survey of 1000 marathon runners, 425 drink Pepsi everyday, 703 drink Gatorade everyday, and 288 drink both of these everyday.

    i) How many runners drink either Pepsi or Gatorade (or both)?

    ii) How many runners drink Pepsi, but not Gatorade?

    iii) How many runners drink neither Pepsi nor Gatorade?

9. The probability that a southern pine beetle infects a Virginia pine tree is 0.17. The probability that an engraver beetle infects a Virginia pine is 0.11. The probability that both of these beetles infect a Virginia pine is 0.05. What is the

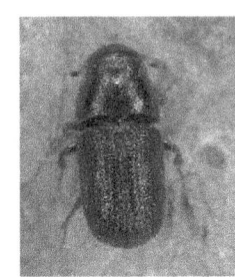

probability of a Virginia pine being infected by neither the southern pine beetle nor the engraver beetle?

10. You are playing roulette at a fair wheel with 38 slots numbered 0, 00, and 1 to 36. The positive even- numbered slots are red in color, the odd-numbered slots are black, and 0 and 00 are green. What are the probabilities of the following events associated with a spin of the wheel?
    i) The ball lands on an odd number.
    ii) The ball lands on a red slot or a green slot.
    iii) The ball lands on numbers 1 through 12 or a red slot.

11. Draw a Punnet square to help answer each of the following questions. In humans, left-handedness appears to be recessive allele to the dominant right-handedness. Suppose Lars is a right-handed person has a parent who is left-handed. (i.e. his genotype has one of each allele.)
i) If Lars mates with a left-handed person, what is the probability they produce a left-handed child?
ii) If Lars mates with a right-handed person who has a left-handed parent, what is the probability they produce a left-handed child?

12. Each week the state lottery pays out $500,000 to the person who has chosen the same 7-digit number that is randomly drawn from a barrel with two million such numbers in it. If each lottery ticket (with one number on it) costs $2, what is your expected value if you purchase one ticket?

Solutions to Odd-Numbered Exercises

Section 7.1

1. i) $S = \{2, 3, 4, 5, 6, 7, 8, 9, 10, 11, 12, 13, 14, 15, 16, 17, 18\}$
ii) $E = \{2, 4, 6, 8, 10, 12, 14, 16, 18\}$
iii) $A = \{10, 11, 12, 13, 14, 15, 16, 17, 18\}$
iv) $L = \{2, 3, 4, 5\}$
v) $P = \{2, 3, 5, 7, 11, 13, 17\}$

3. i) $\{(1, 1), (1, 2), (1, 3), (1, 4), (1, 5), (1, 6),$
      $(2, 1), (2, 2), (2, 3), (2, 4), (2, 5), (2, 6),$
      $(3, 1), (3, 2), (3, 3), (3, 4), (3, 5), (3, 6),$
      $(4, 1), (4, 2), (4, 3), (4, 4), (4, 5), (4, 6)\}$ (Red die listed first in each pair.)
ii) $F = \{(1, 1), (1, 3), (1, 5), (3, 1), (3, 3), (3, 5)\}$
iii) $B = \{(3, 6), (4, 5), (4, 6)\}$
iv) $N = \{((2, 1), (3, 1), (3, 2), (4, 1), (4, 2), (4, 3)\}$
v) $T = \{(1, 3), (2, 4), (3, 5), (4, 6)\}$
vi) $A = \{(1, 3), (1, 4), (1, 5), (1, 6), (2, 4), (2, 5), (2, 6), (3, 5), (3, 6), (4, 6)\}$

5. i) $S = \{0, 1, 2, 3, \ldots, 98, 99, 100\}$
ii) $S = \{10000, 10001, 10002, \ldots 99997, 99998, 99999\}$
iii) $S = \{0, 1, 2, 3\}$
iv) $S = \{HHH, HHT, HTH, HTT, THH, THT, TTH, TTT\}$

7. i) $S = \{HHH, HHT, HTH, HTT, THH, THT, TTH, TTT\}$
ii) $A = \{HHH, HHT, HTH, HTT, THH, THT, TTH\}$
iii) $B = \{HHT, HTH, THH\}$
iv) $C = \{HTT, THT, TTH\}$
v) $D = \{TTT\}$

9. i) $S = \{0, 00, 1, 2, 3, \ldots, 34, 35, 36\}$
ii) $R = \{9, 32, 30, 7, 5, 34, 3, 36, 1, 27, 25, 12, 19, 18, 21, 16, 23, 14\}$ (Ordered clockwise)
iii) $B = \{20, 28, 11, 33, 17, 22, 15, 24, 13, 10, 29, 8, 31, 6, 26, 4, 35, 2\}$ (Ordered clockwise)
iv) $M = \{19, 21, 23, 25, 27, 30, 32, 34\}$
v) $G = \{0, 00\}$
vi) $E = \{2, 4, 6, 8, 10, 12, 14, 16, 18, 20, 22, 24, 26, 28, 30, 32, 34, 36, 0, 00\}$

Section 7.2

1. i) False; ii) True; iii) True; iv) True; v) False; vi) True; vii) False; viii) True
3. i) People who like basketball or hockey; ii) People who do not like football; iii) People who do not like basketball but do like football; iv) People who like neither hockey nor football.
5. i) $\overline{B} = \{-3, -2, 0, 3\}$;   ii) $A \cap B = \{-1, 1\}$;
iii) $\overline{A} \cup B = \{-3, 2, 3\}$;   iv) $\overline{A \cup B} = \{-3, 3\}$;   v) $\overline{A \cup \overline{B}} = \{2\}$
7.
9. i) $\overline{F \cap \overline{S}} = \{1, 3, 4, 6, 8\}$ ii) $\overline{F} \cup \overline{S} = \{1, 2, 3, 5, 6, 7\}$ iii) $F \cap S = \{4, 8\}$ and $F \cap \overline{S} =$

$\{2, 5, 7\}$. So $(F \cap S) \cup (F \cap \overline{S}) = F$; iv) $S \cap F = \{4, 8\}$ and $S \cap \overline{F} = \{3\}$. So
$(S \cap F) \cup (S \cap \overline{F}) = S$.

11. For example, partition them by height: less than 5 ft, 6 in and greater than or equal to 5 ft, 6 in.

15.

17. 19

19. $n(A \cap B) = 7$, $n(B) = 60$, and $n(\overline{B}) = 65$.

21. $g(A) = 2^{n(A)}$; 2^{52}

23.   i) $f(L) = \{\beta, \delta, \gamma\}$ ii) $f(L \cap M) = \varnothing$ iii) $n(L) = 4$
   iv) $n(f(L)) = 3$ v) $n(f(L \cap M)) = 0$

25. i) 845; ii) 205; iii) 155

27. i) 525; ii) 270; iii) 75

29. i) 6; ii) 16; iii) 73; iv) 39; v) 50

31. i) $A \cap B \cap C$; ii) $A \cap (\overline{B \cup C})$; iii) $B \cap (\overline{A \cup C})$; iv) $(A \cup B) \cap \overline{C}$; v) $(B \cap C) \cap \overline{A}$

Section 7.3

1. i) $S = \{HH, HT, TH, TT\}$; ii) 3/4; iii) 1/4; iv) 1/2

3.   i) $\{(1, 1), (1, 2), (1, 3), \ldots, (6, 4), (6, 5), (6, 6)\}$;
   ii) 5/12;   iii) 1/6;   iv) 1/9;   v) 5/18

5. If A and B are mutually exclusive, then $p(A \cap B) = 0$. Therefore, $p(A \cup B) = p(A) + p(B)$.

7.   i) $S = \{10, 11, 12, 13, 14, 15, 16, 17, 18, 19, 20\}$
   ii) $\dfrac{6}{11}$ iii) $\dfrac{5}{11}$ iv) $\dfrac{4}{11}$ v) $\dfrac{8}{11}$

9. i) $\dfrac{1}{2}$ ii) $\dfrac{1}{4}$ iii) $\dfrac{7}{8}$ iv) $\dfrac{1}{8}$ v) $\dfrac{5}{8}$

11. $p(A \cup B \cup C) = 1$ because $U = A \cup B \cup C$. Since A and $B \cup C$ are mutually exclusive, $p(A \cup B \cup C) = p(A) + p(B \cup C)$. Since B and C are mutually exclusive, $p(B \cup C) = p(B) + p(C)$. Therefore, $1 = p(A \cup B \cup C) = p(A) + p(B) + p(C)$.

13. i) $\dfrac{1}{10}$ ii) $\dfrac{1}{100}$ iii) $\dfrac{1}{2}$ iv) $\dfrac{9}{10}$

15. i) $\dfrac{1}{4}$ ii) $\dfrac{3}{4}$ iii) $\dfrac{13}{16}$ iv) $\dfrac{3}{8}$

17. i) $\dfrac{21}{32}$ ii) $\dfrac{27}{32}$ iii) $\dfrac{7}{32}$ iv) $\dfrac{3}{16}$

19. i) 0.845   ii) 0.205   iii) 0.155

21. i) $\dfrac{9}{19}$ ii) $\dfrac{1}{19}$ iii) $\dfrac{11}{38}$ iv) $\dfrac{4}{19}$ v) $\dfrac{10}{19}$

23. –$1.90

25. 0.5

	w	w
R	Rw	Rw
w	ww	ww

27. Yes.   0.25

	B	b
B	BB	Bb
b	Bb	bb

29. Yes. 0.25

	D	d
D	DD	Dd
d	Dd	dd

31. Six spots. The probability is 16/36.
33. 10:3 ; 1:1 ; 2:7

Solutions to Chapter Review Test

1. i) True; ii) True; iii) False; iv) False; v) True; vi) False; vii) True; viii) False
2. No. The given set does not contain all the possible outcomes.
3. i) $\{t, w, x, z\}$; ii) $\{u, y\}$; iii) $\{s, v\}$; iv) $\{t, u, w, x, y, z\}$; v) 5; vi) 3; vii) 1
4. See Figure 7.2.4.
5. 25
6. 250
7. i) 1/2; ii) 1/4; iii) 7/8; iv) 1/8; v) 5/8
8. i) 840; ii) 137; iii) 160
9. 0.77
10. i) 9/19; ii) 10/19; iii) 12/19
11. i) 1/2; ii) 1/4
12. −$1.75

www.ingramcontent.com/pod-product-compliance
Lightning Source LLC
Chambersburg PA
CBHW082151230526
45467CB00044B/2877